The authors' objective has been to concentrate on amino acids and peptides without detailed discussions of proteins, although the book gives all the essential background chemistry, including sequence determination, synthesis and spectroscopic methods, to allow the reader to appreciate protein behaviour at the molecular level. The approach is intended to encourage the reader to cross classical boundaries, such as in the later chapter on the biological roles of amino acids and the design of peptide-based drugs. For example, there is a section on enzyme-catalysed synthesis of peptides, an area often neglected in texts describing peptide synthesis.

This modern text will be of value to advanced undergraduates, graduate students and research workers in the amino acid, peptide and protein field.

Amino Acids and Peptides

Amino Acids and Peptides

G. C. BARRETT

AND

D. T. ELMORE

CAMBRIDGE
UNIVERSITY PRESS

PUBLISHED BY THE PRESS SYNDICATE OF THE UNIVERSITY OF CAMBRIDGE
The Pitt Building, Trumpington Street, Cambridge CB2 1RP, United Kingdom

CAMBRIDGE UNIVERSITY PRESS
The Edinburgh Building, Cambridge CB2 2RU, United Kingdom
40 West 20th Street, New York, NY 10011-4211, USA
10 Stamford Road, Oakleigh, Melbourne 3166, Australia

First published 1998

Printed in the United Kingdom at the University Press, Cambridge

Typeset in Times NR 10/13pt   [SE]

A catalogue record for this book is available from the British Library

ISBN 0 521 46292 4 hardback
ISBN 0 521 46827 2 paperback

Contents

Foreword

This is an undergraduate and introductory postgraduate textbook that gives information on amino acids and peptides, and is intended to be self-sufficient in all the organic and analytical chemistry fundamentals. It is aimed at students of chemistry, and allied areas. Suggestions for supplementary reading are provided, so that topic areas that are not covered in depth in this book may be followed up by readers with particular study interests.

A particular objective has been to concentrate on amino acids and peptides, as the title of the book implies; the exclusion of detailed discussion of proteins is deliberate, but the book gives all the essential background chemistry so that protein behaviour at the molecular level can be appreciated.

There is an emphasis on the uses of amino acids and peptides, and on their biological roles and, while Chapter 8 concentrates on this, a scattering of items of information of this type will be found throughout the book. Important pharmaceutical developments in recent years underline the continuing importance and potency of amino acids and peptides in medicine and the flavour of current research themes in this area can be gained from Chapter 9.

Supplementary reading
(see also lists at the end of each Chapter)

Standard Student Texts

Standard undergraduate Biochemistry textbooks relate the general field to the coverage of this book. Several such topic areas are covered in

Zubay, G. (1993) *Biochemistry*, Third Edition, Wm. C. Brown Communications Inc, Dubuque, IA
and
Voet, D. and Voet, J. G. (1995) *Biochemistry*, Second edition, Wiley, New York

Typically, these topic areas as covered by Zubay are

Chapter 3: 'The building blocks of proteins: amino acids, peptides and proteins'
Chapter 4: 'The three-dimensional structure of proteins'
Chapter 5: 'Functional diversity of proteins'

Removed more towards biochemical themes, are

Chapter 18: 'Biosynthesis of amino acids'
Chapter 19: 'The metabolic fate of amino acids'
Chapter 29: 'Protein synthesis, targeting, and turnover'

Voet and Voet give similar coverage in

Chapter 24: 'Amino acid metabolism'
Chapter 30: 'Translation' (i.e. protein biosynthesis)
Chapter 34: 'Molecular physiology' (of particular relevance to coverage in this book of blood clotting, peptide hormones and neurotransmitters)

Supplementary reading:
suggestions for further reading

(a) Protein structure

Branden, C., and Tooze, J. (1991) *Introduction to Protein Structure*, Garland Publishing Inc., New York

(b) Protein chemistry

Hugli, T. E. (1989) *Techniques of Protein Chemistry*, Academic Press, San Diego, California
Cherry, J. P. and Barford, R. A. (1988) *Methods for Protein Analysis,* American Oil Chemists' Society, Champaign, Illinois

(c) Amino acids

Barrett, G. C., Ed. (1985) *Chemistry and Biochemistry of the Amino Acids*, Chapman and Hall, London
Barrett, G. C. (1993) in *Second Supplements to the 2nd Edition of Rodd's Chemistry of Carbon Compounds*, Volume 1, Part D: Dihydric alcohols, their oxidation products and derivatives, Ed. Sainsbury, M., Elsevier, Amsterdam, pp. 117–66
Barrett, G. C. (1995) in *Amino Acids, Peptides, and Proteins*, A Specialist Periodical Report of The Royal Society of Chemistry, Vol. 26, Ed. Davies, J. S., Royal Society of Chemistry, London (preceding volumes cover the literature on amino acids, back to 1969 (Volume 1))
Coppola, G. M. and Schuster, H. F. (1987) *Asymmetric Synthesis: Construction of Chiral Molecules using Amino Acids*, Wiley, New York
Dawson, R. M. C., Elliott, D. C., Elliott, W. H., and Jones, K. M. (1986) *Data for Biochemical Research*, Oxford University Press, Oxford

Greenstein, J. P., and Winitz, M. (1961) *Chemistry of the Amino Acids*, Wiley, New York (a facsimile version (1986) of this three-volume set has been made available by Robert E. Krieger Publishing Inc., Malabar, Florida)

Williams, R. M. (1989) *Synthesis of Optically Active α-Amino Acids*, Pergamon Press, Oxford

(d) Peptides

Bailey, P. D. (1990) *An Introduction to Peptide Chemistry*, Wiley, Chichester

Bodanszky, M. (1988) *Peptide Chemistry: A Practical Handbook*. Springer-Verlag, Berlin

Bodanszky, M. (1993) *Principles of Peptide Synthesis*, Second Edition, Springer-Verlag, Heidelberg

Elmore, D. T. (1993) in *Second Supplements to the 2nd Edition of Rodd's Chemistry of Carbon Compounds*, Volume 1, Part D: Dihydric alcohols, their oxidation products and derivatives, Ed. Sainsbury, M., Elsevier, Amsterdam, pp. 167–211

Elmore, D. T. (1995) in *Amino Acids, Peptides, and Proteins*, A Specialist Periodical Report of The Royal Society of Chemistry, Vol. 26, Ed. Davies, J. S., Royal Society of Chemistry, London (preceding volumes cover the literature of peptide chemistry back to 1969 (Volume 1))

Jones, J. H. (1991) *The Chemical Synthesis of Peptides*, Clarendon Press, Oxford

1

Introduction

1.1 Sources and roles of amino acids and peptides

More than 700 amino acids have been discovered in Nature and most of them are α-amino acids. Bacteria, fungi and algae and other plants provide nearly all these, which exist either in the free form or bound up into larger molecules (as constituents of peptides and proteins and other types of amide, and of alkylated and esterified structures).

The twenty amino acids (actually, nineteen α-amino acids and one α-imino acid) that are utilised in living cells for protein synthesis under the control of genes are in a special category since they are fundamental to all life forms as building blocks for peptides and proteins. However, the reasons why all the other natural amino acids are located where they are, are rarely known, although this is an area of much speculation. For example, some unusual amino acids are present in many seeds and are not needed by the mature plant. They deter predators through their toxic or otherwise unpleasant characteristics and in this way are thought to provide a defence strategy to improve the chances of survival for the seed and therefore help to ensure the survival of the plant species.

Peptides and proteins play a wide variety of roles in living organisms and display a range of properties (from the potent hormonal activity of some small peptides to the structural support and protection for the organism shown by insoluble proteins). Some of these roles are illustrated in this book.

1.2 Definitions

The term 'amino acids' is generally understood to refer to the *aminoalkanoic acids*, H_3N^+—$(CR^1R^2)_n$—CO_2^- with $n = 1$ for the series of α-amino acids, $n = 2$ for β-amino acids, etc. The term '*dehydro-amino acids*' specifically describes *2,3-unsaturated (or 'αβ-unsaturated')-2-aminoalkanoic acids*, H_3N^+—$(C=CR^1R^2)$—CO_2^-.

However, the term '*amino acids*' would include all structures carrying amine and acid functional groups, including simple aromatic compounds, e.g. anthranilic acid,

$$m [H_3N^+ -(R^1R^2C-)_n CO_2^-]$$

$$\downarrow$$

$$H_3N^+ -(R^1R^2C-)_n CO[-NH-(R^1R^2C-)_n CO]_{m-2} -NH-(R^1R^2C-)_n CO_2^- + (m-1)\ H_2O$$

(various R^1, R^2; various n, m)

Condensation of *m* molecules of an α-amino acid to give one molecule of an oligopeptide containing *m* α-amino acid residues

$$H_3N^+ -R^1\overset{*}{C}H-CO-NH-R^2\overset{*}{C}H-CO-NH-R^3\overset{*}{C}H-CO-NH-R^4\overset{*}{C}H-CO-NH-R^5\overset{*}{C}H-CO_2^-$$

A Pentapeptide [Leu-Enkephalin, Tyr-Gly-Gly-Phe-Leu, when $R^1 = CH_2-C_6H_4-p-OH$; $R^2 = R^3 = H$; $R^4 = CH_2-C_6H_5$; $R^5 = CH_2-CH(CH_3)_2$] with the L-configuration [α(S)-configuration in this particular example] at each chiral centre C*

Figure 1.1. Peptides as condensation polymers of α-amino acids.

o-H_3N^+—C_6H_4—CO_2^-, and would also cover other types of acidic functional groups (such as phosphorus and sulphur oxy-acids, H_3N^+—$(R^1R^2C$—$)_n HPO_3^-$ and R_3N^+—$(R^1R^2C$—$)_n SO_3^-$, etc). The family of boron analogues $R_3N^·BHR^1$—CO_2R^2 (˙ denotes a dative bond) has recently been opened up through the synthesis of some examples (Sutton *et al.*, 1993); it would take only the substitution of the carboxy group in these 'organoboron amino acids' ($R = R^1 = R^2 = H$) by phosphorus or sulphur equivalents to obtain an amino acid that contains no carbon! However, unlike the amino acids containing sulphonic and phosphonic acid groupings, naturally occurring examples of organoboron-based amino acids are not known.

The term *'peptides'* has a more restricted meaning and is therefore a less ambiguous term, since it covers polymers formed by the condensation of the respective amino and carboxy groups of α, β, γ . . . -amino acids. For the structure with $m = 2$ in Figure 1.1 (i.e., for a dipeptide) up to values of $m \simeq 20$ (an eicosapeptide), the term *'oligopeptide'* is used and a prefix *di-, tri-, tetra-, penta-* (see Leu-enkephalin, a linear pentapeptide, in Figure 1.1), . . . *undeca-* (see cyclosporin A, a cyclic undecapeptide, in Figure 1.4 later), *dodeca-*, . . . etc. is used to indicate the number of *amino-acid residues* contained in the compound. *Homodetic* and *heterodetic* peptides are illustrated in Chapter 7.

Isopeptides are isomers in which amide bonds are present that involve the *side-chain amino group* of an αω-di-amino acid (e.g. lysine) or of a poly-amino acid and/or *the side-chain carboxy-group of an α-amino-di- or -poly-acid* (e.g. aspartic acid or glutamic acid). Glutathione (Chapter 8) is a simple example. Longer polymers are termed *'polypeptides'* or *'proteins'* and the term *'polypeptides'* is becoming the most commonly used general family name (though proteins remains the preferred term for particular examples of large polypeptides located in precise biological contexts). Nonetheless, the relationship between these terms is a little more contentious, since the change-over from polypeptide to protein needs definition. The figure 'roughly fifty amino acid residues' is widely accepted for this. Insulin (a polymer of fifty-one α-amino acids but consisting of two crosslinked oligopeptide chains; see Figure 1.4 later) is on the borderline and has been referred to both as a *small protein* and as a *large polypeptide*.

Poly(α-amino acid)s is a better term for peptides formed by the self-condensation of one amino acid; natural examples exist, such as poly(D-glutamic acid), the protein coat of the anthrax spore (Hanby and Rydon, 1946). In early research in the textile industry, poly(α-amino acid)s showed promise as synthetic fibres, but the synthesis methodology required for the polymerisation of amino acids was complex and uneconomic.

Polymers of controlled structures made from N-alkyl-α-amino acids (Figure 1.1; —NR^n instead of —NH—, $R^1 = R^2 = H$; $n = 1$), i.e. $H_2^+NR^n$—CH_2CO—$[NR^n$— CH_2—CO—$]_mNR^n$—CH_2—CO_2^-, which are poly(N-alkylglycine)s of defined sequence (various R^n at chosen points along the chain), have been synthesised as *peptide mimetics* (see Chapter 9) and have been given the name *peptoids*. These can be viewed as peptides with side-chains shifted from carbon to nitrogen; they will therefore have a very different conformational flexibility (see Chapter 2) from that of peptides and will also be incapable of hydrogen bonding. This is a simple enough way of providing all the correct side-chains on a flexible chain of atoms, in order to mimic a biologically active peptide, but the mimic can avoid enzymic breakdown before it reaches the site in the body where it is needed.

Using the language of polymer chemistry, polypeptides made from two or more different α-amino acids are *copolymers* or irregular poly(amide)s, whereas poly(α-amino acid)s, H—$[NH$—CR^1R^2—CO—$]_mOH$, are *homopolymers* that could be described as members of the nylon[2] family.

Depsipeptides are near-relatives of peptides, with one or more *amide bonds replaced by ester bonds*; in other words, they are formed by condensing α-amino acids with α-hydroxy-acids in various proportions. There are several important natural examples of these, of defined sequence; for example the antibiotic valinomycin and the family of enniatin antibiotics. Structures of other examples of depsipeptides are given in Section 4.8.

Nomenclature for conformational features of peptide structure is covered in Chapter 2.

1.3 'Protein amino acids', alias 'the coded amino acids'

The twenty L-amino acids (actually, nineteen α-amino acids and one α-imino acid (Table 1.1)) which, in preparation for their role in protein synthesis, are joined *in vivo* through their carboxy group to tRNA to form α-aminoacyl-tRNAs, are organised by ribosomal action into specific sequences in accordance with the genetic code (Chapter 8).

'Coded amino acids' is a better name for these twenty amino acids, rather than 'protein amino acids' or 'primary protein amino acids' (the term 'coded amino acids' is increasingly used), because changes can occur to amino-acid residues after they have been laid in place in a polypeptide by ribosomal synthesis. Greenstein and

Table 1.1. *The twenty 'coded' amino acids (nineteen 'coded' L-α-amino acids, and one 'coded' L-α-imino acid): structures and definitions*[a]

Structure conventions for the L-α-amino acids

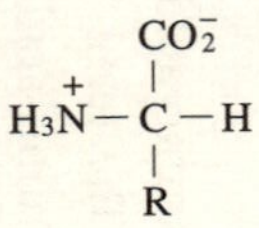

is equivalent to

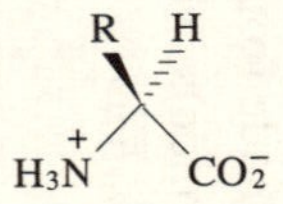

which is equivalent to

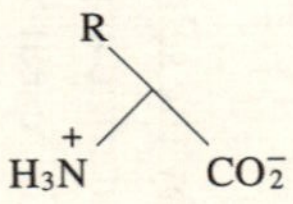

Barrett representation of an L-α-amino acid

Fischer projection of an L-α-amino acid, requiring the carbon chain to be arranged vertically, with the carboxy group at the top

One of the commonly-used three-dimensional representations of an L-α-amino acid

Name of amino acid	Three-letter abbreviation	Single-letter abbreviation	Structures		Hydrophobicity		Hydrophilicity
				*Amino acid side-chain, R =	High	High	High
One with no side-chain* (i.e. with a hydrogen atom)							
Glycine	Gly	G		H			*
Four with saturated aliphatic side-chains* (hydrophobic side-chains)							
Alanine	Ala	A		CH₃		*	
Leucine	Leu	L		CH₂CH(CH₃)₂		*	
Valine	Val	V		CH(CH₃)₂		*	
Isoleucine	Ile	I		(S)-CH(CH₃)C₂H₅		*	

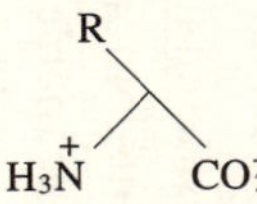

Group	Name	Three-letter	One-letter	Side-chain, R
Ten with functionalised aliphatic side-chains* (mostly hydrophilic side-chains)	Arginine	Arg	R	$CH_2CH_2CH_2NHC(=NH)NH_2$
	Aspartic acid	Asp	D	CH_2CO_2H
	Asparagine	Asn	N	CH_2CONH_2
	Glutamic acid	Glu	E	$CH_2CH_2CO_2H$
	Glutamine	Gln	Q	$CH_2CH_2CONH_2$
	Lysine	Lys	K	$CH_2CH_2CH_2CH_2NH_2$
	Methionine	Met	M	$CH_2CH_2SCH_3$
	Cysteine	Cys	C	CH_2SH
	Serine	Ser	S	CH_2OH
	Threonine	Thr	T	$(R)\text{-}CH(CH_3)OH$
Four with aromatic or heteroaromatic side-chains* (most of these side-chains are hydrophobic)	Phenylalanine	Phe	F	$CH_2C_6H_5$
	Tyrosine	Tyr	Y	$CH_2\text{-}(p\text{-}OH\text{-}C_6H_4)$
	Histidine	His	H	$CH_2\text{-}(imidazol\text{-}4\text{-}yl)$
	Tryptophan	Trp	W	$CH_2\text{-}(indol\text{-}3\text{-}yl)$
The 'coded' α-imino acid	Proline	Pro	P	

Notes:

1. The structure of each side-chain, R, is given for the 19 'coded α-amino acids', after each name. The full structure of the 'coded α-imino acid' proline is given. 'Three-letter' and 'one-letter' abbreviations are given for all twenty. The *three-letter* abbreviation is the *first three letters of the name* for all twenty, *except* for asparagine (Asn), glutamine (Gln), isoleucine (Ile) and tryptophan (Trp). The *single-letter* abbreviated name is the first letter of their full name for *eleven* of them. Different letters are needed for the *other nine*, to avoid ambiguity: arginine (R), asparagine (N), aspartic acid (D), glutamic acid (E), glutamine (Q), lysine (K), phenylalanine (F), tryptophan (W) and tyrosine (Y).

2. All full names end in 'ine' except aspartic acid, glutamic acid and tryptophan. Adjectives are derived from the names by dropping the 'ine' or its equivalent ending and adding 'yl'; thus, alanyl, glutamyl, prolyl, tryptophyl, etc.

3. *Configurations*. The 'R/S' convention can easily be transferred to replace the Fischer 'D/L' system, while retaining the trivial names: L-enantiomers of all the coded amino acids are members of the S series except L-cysteine, which becomes R-cysteine through proper application of the R/S rules. Diastereoisomers (the isoleucine/allo-isoleucine and threonine/allothreonine pairs, 'allo' indicating inversion of the side-chain configuration of the coded amino acid) are less ambiguously named through the 'R/S' system, although the side-chain configuration can be indicated; for example, natural L-isoleucine is (2S,3S)-isoleucine:

Table 1.1. (*cont.*)

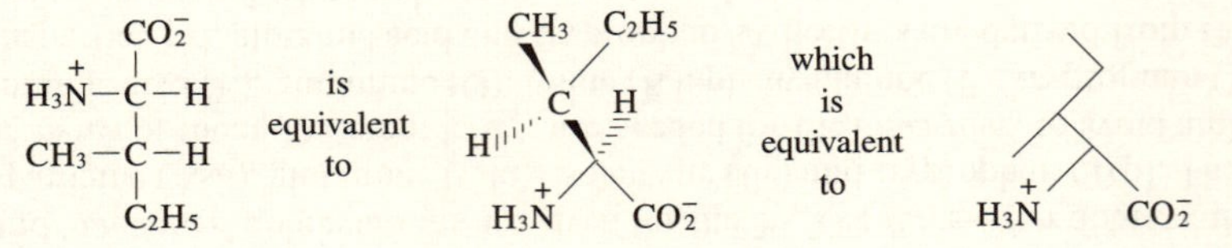

$$\underset{\underset{\displaystyle C_2H_5}{\overset{\displaystyle |}{CH_3-\overset{\displaystyle H}{\underset{\displaystyle |}{C}}-H}}}{\overset{\displaystyle CO_2^-}{\overset{\displaystyle |}{\underset{+}{H_3N}-C-H}}} \quad \text{is equivalent to} \quad \overset{CH_3 \quad C_2H_5}{\underset{\underset{+}{H_3N}}{\overset{|}{C}}\;\overset{H}{\underset{CO_2^-}{}}}$$

whereas L-*alloisoleucine* is (2S,3R)-isoleucine:

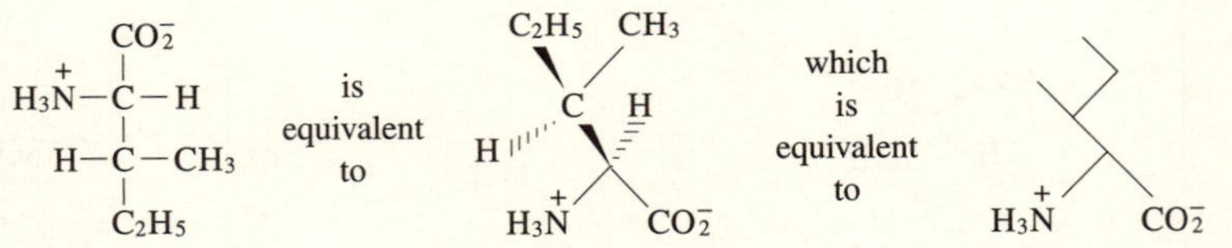

$$\underset{\underset{\displaystyle C_2H_5}{\overset{\displaystyle |}{H-\overset{\displaystyle CH_3}{\underset{\displaystyle |}{C}}}}}{\overset{\displaystyle CO_2^-}{\overset{\displaystyle |}{\underset{+}{H_3N}-C-H}}} \quad \text{is equivalent to} \quad \overset{C_2H_5 \quad CH_3}{\underset{\underset{+}{H_3N}}{\overset{|}{C}}\;\overset{H}{\underset{CO_2^-}{}}}$$

For the structures of natural L-*threonine* ((2S,3R)-threonine) and L-*allothreonine* ((2S,3S)-threonine), replace the side-chain ethyl group (C_2H_5) in isoleucine and alloisoleucine by OH.

4. *IUPAC–IUB nomenclature recommendations* (1983), reproduced in full in *Amino Acids, Peptides, and Proteins*, 1985, Vol. 16, The Royal Society of Chemistry, p. 387; and in *Eur.J.Biochem.*, 1984, **138**, 9, encourage the retention of trivial names for the common α-amino acids, but systematic names are relatively straightforward; thus, L-alanine is 2S-aminopropanoic acid and L-histidine is 2S-amino-3-(imidazol-4-yl)-propanoic acid (the name for the predominant tautomer).

5. 'Hydrophilic' and 'hydrophobic' are terms used to denote the relative water-attracting and water-repelling property, respectively, of the side-chain when the amino acid is condensed into a polypeptide (see Chapter 5). The term 'hydropathy index' may be used to place the amino acids in order of their 'hydrophilicity' (Kyte and Doolittle, 1985), and their relative positions are shown here on an arbitrary scale.

[a] Selenocysteine (i.e. cysteine with the sulphur atom replaced by a selenium atom) has been found in certain proteins, e.g. formate dehydrogenase, an enzyme from *Escherichia coli*, and it has very recently been shown to be placed there through normal ribosomal synthesis (Stadtman, 1996). Thus selenocysteine can now be accepted as the 'twenty-first coded amino acid'.

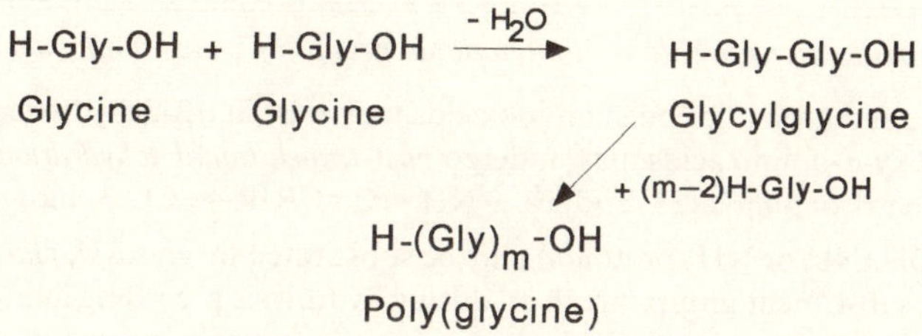

H-Gly-OH + H-Gly-OH $\xrightarrow{\;-\,H_2O\;}$ H-Gly-Gly-OH

Glycine   Glycine   Glycylglycine

$+\,(m-2)\,$H-Gly-OH

H-(Gly)$_m$-OH

Poly(glycine)

Figure 1.2. Polymerisation of glycine.

Winitz, in their 1961 book, listed 'the 26 protein amino acids', six of which were later found to be formed from among the other twenty 'protein amino acids' in the list of Greenstein and Winitz, after the protein had left the gene ('*post-translational* (sometimes called *post-ribosomal*) *modification*' or '*post-translational processing*'). Because of these changes made to the polypeptide after ribosomal synthesis, amino acids that are not capable of being incorporated into proteins by genes ('secondary protein amino acids', Table 1.2) can, nevertheless, be found in proteins.

1.4 Nomenclature for 'the protein amino acids', alias 'the coded amino acids'

The common amino acids are referred to through trivial names (for example, glycine would not be named either 2-aminoethanoic acid or amino-acetic acid in the amino acid and peptide literature). Table 1.1 summarises conventions and gives structures. The rarer natural amino acids are usually named as derivatives of the common amino acids, if they do not have their own trivial names related to their natural source (Table 1.2), but apart from these, there are occasional examples of the use of systematic names for natural amino acids.

1.5 Abbreviations for names of amino acids and the use of these abbreviations to give names to polypeptides

To keep names of amino acids and peptides to manageable proportions, there are agreed conventions for nomenclature (see the footnotes to Table 1.1). The simplest α-amino acid, glycine, would be depicted H—Gly—OH in the standard 'three-letter' system, the H— and —OH representing the 'H$_2$O' that is expelled when this amino acid undergoes condensation to form a peptide (Figure 1.2). The three-letter abbreviations therefore represent the 'amino-acid residues' that make up peptides and proteins.

So this '*three-letter system*' was introduced, more with the purpose of space-saving nomenclature for peptides than to simplify the names of the amino acids. A '*one-letter system*' (thus, glycine is G) is more widely used now for peptides (but is never used to refer to individual amino acids in other contexts) and is restricted to naming peptides synthesised from the coded amino acids (Figure 1.3).

7

Table 1.2. *Post-translational changes to proteins: the modified coded amino acids present in proteins, including crosslinking amino acids (secondary amino acids)*

Modifications to side-chain functional groups of coded amino acids

1. The aliphatic and aromatic coded amino acids may exist in αβ-dehydrogenated forms and the β-hydroxy-α-amino acids may undergo *post-translational dehydration*, so as to introduce αβ-dehydroamino acid residues, $-NH-(C=CR^1R^2)-CO-$, into polypeptides.

2. Side-chain OH, NH or NH_2 proton(s) may be substituted by *glycosyl, phosphate* or *sulphate*. These substituent groups are 'lost' during hydrolysis preceding analysis and during laboratory treatment of proteins by hydrolysis prior to chemical sequencing, which creates a problem that is usually solved through spectroscopic and other analytical techniques.

3. Side-chain NH_2 of lysine may be *methylated* or *acylated*: (N^ε-methylalanyl, N^ε-di-aminopimelyl).

4. Side-chain NH_2 of glutamine may be *methylated*; giving N^5-methylglutamine, and the side-chain NH_2 of asparagine may be *glycosylated*.

5. Side-chain CH_2 may be *hydroxylated*, e.g. hydroxylysine, hydroxyprolines (trans-4-hydroxyproline in particular), or *carboxylated*, e.g. to give α-aminomalonic acid, β-carboxyaspartic acid, γ-carboxyglutamic acid, β-hydroxyaspartic acid, etc.

6. Side-chain aromatic or heteroaromatic moieties may be *hydroxylated, halogenated or N-methylated.*

7. The side-chain of arginine may be modified (e.g. to give ornithine (Orn), $R=CH_2CH_2CH_2NH_2$, or citrulline (Cit), $R=CH_2CH_2CH_2NHCONH_2$).

8. The side-chain of cysteine may be modified, as in 1 above, also selenocysteine (CH_2SeH instead of CH_2SH; see footnote a to Table 1.1), lanthionine (see 10 below).

9. The side-chain of methionine may be S-alkylated (see Table 1.3) or oxidised at S to give methionine sulphoxide.

10. Crosslinks in proteins may be formed by condensation between nearby side-chains.
    (a) From lysine: e.g. lysinoalanine as if from [lysine+serine$-H_2O$]

    $$\rightarrow \quad \text{dehydroalanine} \quad \rightarrow \quad \begin{matrix} \text{H-Lys-OH} \\ | \\ \text{H-Ala-OH} \end{matrix}$$

    (b) From tyrosine: 3,3′-dityrosine, 3,3′,5′,3″-tertyrosine, etc.
    (c) From cysteine: oxidation of the thiol grouping ($HS-+-SH\rightarrow-S-S-$) to give the disulphide or to give cysteic acid (Cya): $-SH\rightarrow-SO_3H$ and alkylation leading to sulphide formation (e.g. alkylation as if by dehydroalanine to give lanthionine):

    $$2H-Cys-OH \;\rightarrow\; \overset{\displaystyle \overline{\quad\quad S \quad\quad}}{H-Ala-OH \quad H-Ala-OH}$$

    (Further examples of crosslinking amino acids in peptides and proteins are given in Section 5.11.)

Nomenclature of post-translationally modified amino acids

Abbreviated names for close relatives of the 'coded amino acids' can be based on the 'three-letter' names when appropriate; thus, L-Pro after post-translational hydroxylation gives L-Hypro (trans-4-hydroxyproline, or (2S,4R)-hydroxyproline).

Current nomenclature recommendations (see footnote to Table 1.1) allow a number of abbreviations to be used for some non-coded amino acids possessing trivial names (some of which are used above and elsewhere in this book): Dopa, β-Ala, Glp, Sar, Cya, Hcy (homocysteine) and Hse (homoserine) are among the more common.

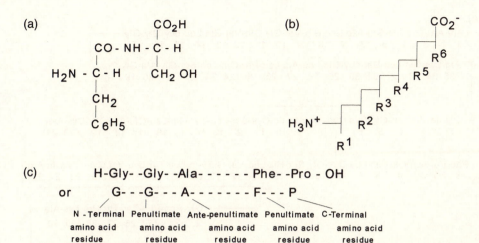

Figure 1.3. (a) The dipeptide L-phenylalanyl-L-serine in the Fischer depiction. (b) The schematic structure of a hexapeptide in the Fischer depiction, resulting in inefficient use of space. (c) The 'three-letter' and 'one-letter' conventions for a representative peptide, GGA---FP.

The 'three-letter system' has some advantages and has gradually been extended (Figure 1.4) to encompass several amino acids other than the coded amino acids. It is usually used to display schemes of laboratory peptide synthesis (Chapter 7) since it allows protecting groups and other structural details to be added, something that is very difficult and often confusing if attempted with the one-letter system.

The one-letter abbreviation (like its three-letter equivalent) represents 'an amino-acid residue' and the system allows the structure of a peptide or protein to be conveniently stated as a string of letters, written as a line of text, incorporating the long-used convention that the amino terminus (the 'N-terminus') is to the LEFT and the carboxy terminus (the 'C-terminus') is to the RIGHT. This convention originates in the Fischer projection formula for an L-α-amino acid or a peptide made up of L-α-amino acids; the L-configuration places the amino group to the left and the carboxy group to the right in a structural formula, as in Figure 1.3.

There are increasing numbers of violations of these rules; N-acetyl alanine, for example, being likely to be abbreviated Ac—Ala in the research literature or its correct abbreviation Ac—Ala—OH (but never Ac—A). This does not usually lead to ambiguity on the basis of the rule that peptide structures are written with the N-terminus to the left and the C-terminus to the right. Thus, Ac—Ala should still be correctly interpreted by a reader to mean CH_3—CO—NH—$CH(CH_3)$—CO_2H when this rule is kept in mind, since Ala—OAc (more correctly, H—Ala—OAc) would represent the 'mixed anhydride' NH_2—$CH(CH_3)$—CO—O—CO—CH_3 (there is a footnote about the term 'mixed anhydride' on p. 152).

(a)

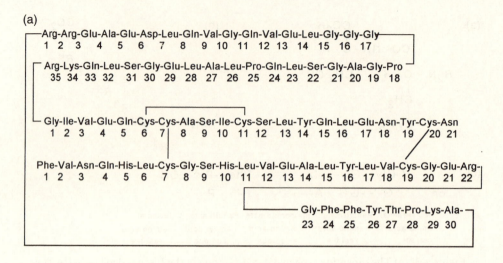

(b)

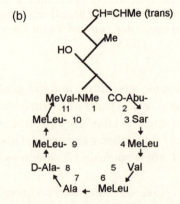

Figure 1.4. Post-translationally modified peptides: (a) Human proinsulin. (b) Cyclosporin A (Me is CH$_3$). As well as the post-translationally modified threonine derivative (residue 1, called 'MeBmt'), cyclosporin A contains one D-amino acid, four N-methyl-L-leucine residues, one 'non-natural' amino acid, Abu (butyrine, side-chain C$_2$H$_5$), Sar (sarcosine, N-methylglycine) and N-methyl-L-valine, but only two of the eleven residues are coded L-amino acids, valine and alanine.

Links through functional groups in side-chains of the amino-acid residues can be indicated in abbreviated structures of peptides (Figure 1.4). Cyclisation between the C- and N-termini to give a cyclic oligopeptide can also be shown in abbreviated structural formulae. Insulin (Figure 1.4) provides an example of the relatively common 'disulphide bridge' (there are three of these in the molecule), whereas cyclosporin A (a cyclic undecapeptide from *Trichoderma inflatum*, which is valuable for its immunosuppressant property that is exploited in organ-transplant surgery) is a product of post-translational cyclisation (Figure 1.4).

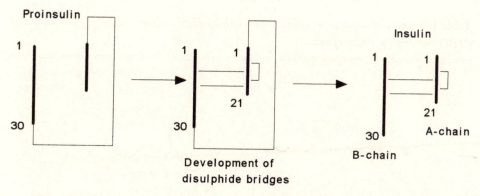

Figure 1.5. Generation of the active hormone, insulin, from the translated peptide, proinsulin (Chan and Steiner, 1977).

1.6 Post-translational processing: modifications of amino-acid residues within polypeptides

The major classes of *structurally altered amino-acid side-chains* within ribosomally synthesised polypeptides, which are achieved by intracellular reactions, are listed in Table 1.2 (see also Chapter 8).

1.7 Post-translational processing: *in vivo* cleavages of the amide backbone of polypeptides

Changes to the amide backbone of the polypeptide through enzymatic cleavages transform the inactive polypeptide into its fully active shortened form. The polypeptide may be transported to the site of action after ribosomal synthesis and then processed there. Standard terminology has emerged for the extended polypeptides, *pre-*, *pro-* and *prepro-peptides* for the inactive *N*-terminal-extended, *C*-terminal-extended and *N*- and *C*-terminal extended forms, respectively, of the active compound. Figure 1.5 shows schematically the post-translational stages from human proinsulin (Figure 1.4) to insulin.

1.8 'Non-protein amino acids', alias 'non-proteinogenic amino acids' or 'non-coded amino acids'

This further term is needed since there are several examples of higher organisms that utilise 'non-protein α-amino acids' that are available in cells in the free form (α-amino acids that are normally incapable of being used in ribosomal synthesis). Some of these free amino acids (Table 1.3) play important roles, one example being *S*-adenosyl-L-methionine, which is a 'supplier of cellular methyl groups'; for example, for the biosynthesis of neuroactive amines (and also for the biosynthesis of many

11

Table 1.3. *Some non-protein amino acids with biological roles, that are excluded from ribosomal protein synthesis*

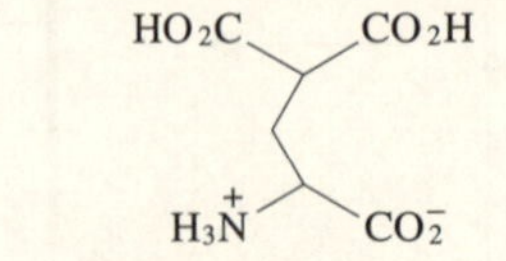

γ-Carboxy-L-glutamic acid[1]

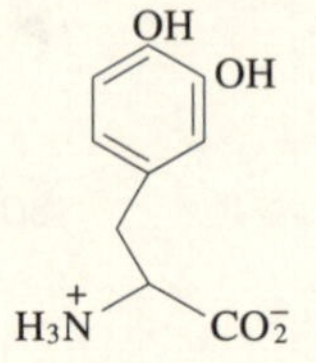

L-Dopa[2]

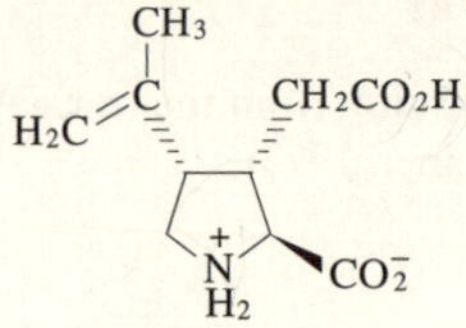

Kainic acid[3]
(3S)-Carboxymethyl-(4S)-
isopropenyl-(S)-proline

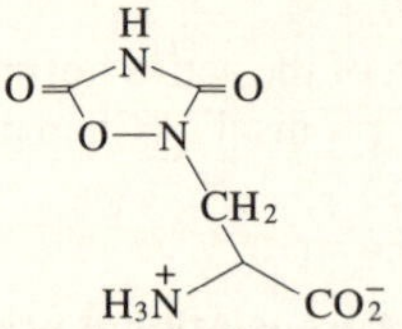

Quisqualic acid[4]

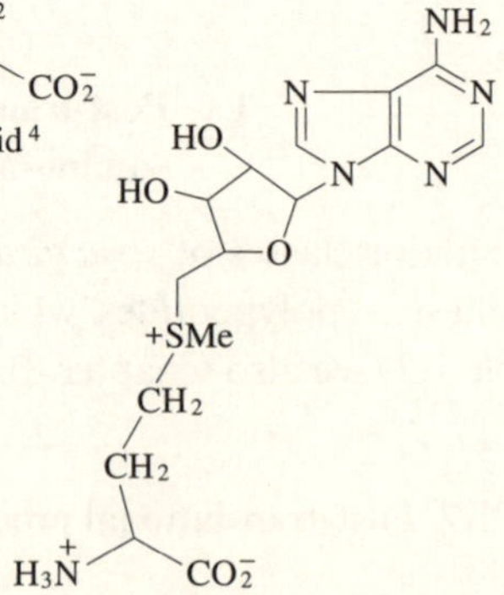

Di-L-tyrosine[5]

S-Adenosyl-L-methionine[6]

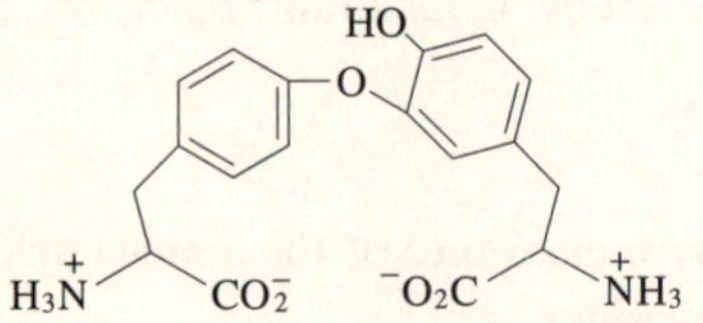

Notes:
1. γ-Carboxyglutamic acid, a constituent of calcium-modulating proteins (introduced through post-translational processing of glutamic acid residues).
2. Treatment for Parkinson's disease.
3. Potent excitatory effects, parent of a family of toxic natural kainoids present in fungi. Domoic acid, which has *trans, trans*-$CHMe-CH=CH-CH=CH-CHMeCO_2H$ in place of the isopropenyl side-chain of kainic acid, is extraordinarily toxic, with fatalities ensuing through eating contaminated shellfish (Baldwin *et al*, 1990).
4. Exhibits potent NMDA receptor activity.
5. One of a number of protein crosslinks.
6. Widely distributed in cells.

other methylated species). Another physiologically important α-amino acid in this category is L-Dopa, the precursor of dopamine in the brain, which is used for treatment of afflictions such as Parkinson's disease and to bring about the return from certain comatose states (described in the book *Awakenings* by Oliver Sacks) that may be induced by L-Dopa.

12

Of course, most of the '700 or so natural amino acids' mentioned at the start of this chapter will be 'non-protein amino acids'. All these were, until recently, thought to be rigorously excluded from protein synthesis and other cellular events that are crucial to life processes, but a very few of these that are structurally related to the coded amino acids may be incorporated into proteins under laboratory conditions. This has been achieved by biosynthesising proteins in media that lack the required coded amino acid, but which contain a close analogue. For example, incorporation of the four-membered-ring α-imino acid azetidine-2-carboxylic acid instead of the five-membered-ring proline and incorporation of norleucine (side-chain $CH_2CH_2CH_2CH_3$) instead of methionine (side-chain $CH_2CH_2SCH_3$) have been demonstrated (Richmond, 1972) and β-(3-thienyl)-alanine has been assimilated into protein synthesis by *E. coli* (Kothakota *et al.*, 1995). Unusual amino acids that are not such close structural relatives of the coded amino acids have been coupled in the laboratory to tRNAs, then shown to be utilised for ribosomal peptide synthesis *in vivo* (Noren *et al.*, 1989).

Ways have been found, in the laboratory, of broadening the specificity of some enzymes (particularly the proteinases, but also certain lipases that can be used in laboratory peptide synthesis; see Chapter 7), for example by employing organic solvents, so that these enzymes catalyse some of the reactions of non-protein amino acid derivatives and some of the reactions of peptides that incorporate unusual amino acids. It has proved possible to involve D-enantiomers of the coded amino acids and D- and L-isomers of non-protein amino acids in peptide synthesis, to generate 'non-natural' peptides.

1.9 Coded amino acids, non-natural amino acids and peptides in nutrition and food science and in human physiology

The nutritional labels for some of the protein amino acids, such as 'essential amino acids', are an indication of their roles in this context. The meaning of the term *'essential'* differs from species to species and reflects the dependence of the organism on certain ingested amino acids that it cannot synthesise for itself, but which it needs in order to be able to generate its life-sustaining proteins. For the human species, the essential amino acids are the L-enantiomers of leucine, valine, isoleucine, lysine, methionine, threonine, phenylalanine, histidine and tryptophan. This implies that the other coded amino acids can be obtained from these essential amino acids, if not through other routes. There are some surprising pathways. For example, cysteine can be generated from methionine, the 'loss' of the side-chain carbon atoms being achieved through passage via cystathionine (Finkelstein, 1990); but homocysteine, the presence of which has been implicated as a causal factor in vascular disease, is also formed first in this route by demethylation of methionine. The D- and L-enantiomers of coded amino acids generally have different tastes and it has recently been appreciated that many fermented foods, such as yoghourt and shellfish (amongst many other food sources), contain substantial amounts of the D enantiomers of the coded amino acids.

The contribution of the D enantiomers to the characteristic taste of foods is currently being evaluated, but it is clear that the D enantiomers generally taste 'sweeter', or at least 'less bitter', than do their L isomers. Of course, kitchen preparation can involve many subtle chemical changes that enhance the attractiveness of natural foodstuffs, including racemisation (Man and Bada, 1987); therefore D enantiomers may be introduced in this way. Peptides are taste contributors, for example the bitter-tasting dipeptides Trp—Phe and Trp—Pro and the tripeptide Leu—Pro—Trp that are formed in beer yeast residues (Matsusita and Ozaki, 1993).

Some coded amino acids are acceptable as food additives and some are widely used in this way (e.g. L-glutamic acid and its monosodium salt). Addition of amino acids to the diet is unnecessary for people already eating an adequate and balanced food supply and the toxicity of even the essential amino acids (methionine is the most toxic of all the coded amino acids (Food and Drugs Administration, Washington USA, 1992)) should be better publicised, because some coded amino acids are easily available (for use in specialised diets by 'body-builders', for example) and are sometimes used unwisely. The use of L-tryptophan for its putative anti-depressant and other 'health' properties was responsible for the outbreak of eosinophilia myalgia syndrome that affected more than 1500 persons (with more than 30 fatalities) in the USA during 1989–90, though the problem was ascribed not to the amino acid itself but rather to an impurity introduced into the amino acid during manufacture (Smith *et al.*, 1991). At the other end of the scale, some amino acids have more trivial uses, e.g. L-tyrosine in sun-tan lotion for cosmetic 'browning' of the skin.

Methionine is included in some proprietary paracetamol products (Pameton; Smith Kline Beecham), since it counteracts some serious side-effects that are encountered with paracetamol overdosing through helping to restore glutathione levels that are the body's natural defence against products of oxidised paracetamol. However, the recommended antidote (bearing in mind the toxicity of methionine) is intravenous *N*-acetyl-L-cysteine, which, in any case, reaches the liver of the overdosed patient faster.

Derivatives of aspartic acid have special importance in neurological research; the *N*-acetyl derivative is a putative marker of neurones and *N*-methyl-D-aspartic acid (NMDA) is creating interest for its possible links with Alzheimer's disease. NMDA is a potent excitant of spinal neurones; there are receptors in the brain for this α-imino acid, for which agonists/antagonists are being sought. A particular interaction being studied is that between ethanol and NMDA receptors (Collingridge and Watkins, 1994; see also Meldrum, 1991).

The industrial production base that has been developed to meet these demands (see Chapter 6) makes many amino acids cheaply available for other purposes such as laboratory use and has contributed in no small measure to the development of the biotechnological sector of the chemical industry.

1.10 The geological and extra-terrestrial distribution of amino acids

The development of sensitive analytical methods for amino acids became an essential support for the study of geological specimens (terrestrial ones and Lunar and Martian samples) from the 1970s. Some of the 'primary and secondary protein amino acids' (and some non-protein amino acids) were established to exist in meteorites (certainly in one of the largest known, the Murchison meteorite from Western Australia) though they have not been found in lunar samples. The scepticism that greeted an inference from this discovery – the inference that life as we know it exists, or once existed, on other planetary bodies – has also boosted interest in the chemistry of the amino acids to try to support alternative explanations for their presence in meteorites. The possibility that such relatively sensitive compounds could have survived the trauma experienced by meteorites penetrating the Earth's atmosphere was soon rejected. They must have been synthesised in the meteorites during the final traumatic stage of their journey. This conclusion was obtained bearing in mind the relevant amino-acid chemistry (Chapter 4); even the common, relatively much more gentle, laboratory practice of ultrasonic treatment of geological and biological samples prior to amino-acid analysis was hastily discouraged when it was found that this causes chemical structural changes to certain common amino acids (e.g. conversion of glutamic acid into glycine); and the injection of energy into mixtures of certain simple compounds also causes the formation of amino acids (Chapter 6).

The use of telescopic spectroscopy has revealed the existence of glycine in interstellar dust clouds. Since these clouds amount to huge masses of matter (greater than the total mass of condensed objects such as stars and planets), there must be universal availability of amino acids, even though they are dispersed thinly in the vast volume of space.

1.11 Amino acids in archaeology and in forensic science

Amino-acid analysis of relatively young fossils and of other archaeological samples has provided information on their age and on the average temperature profiles that characterised the Earth at the time of life of these samples. Samples from living organisms containing protein that has ceased turnover, i.e. proteins in metabolic *culs-de-sac* such as tooth and eye materials, can be analysed for their degree of racemisation of particular amino acids (Asp and Ser particularly; Leu for older specimens) in order to provide this sort of information. The D:L ratio for the aspartic acid present in these sources can be interpreted to assign an age to the organism, since racemisation of this amino acid is relatively rapid on the geological time scale and even in terms of life-span of a human being. The D:L ratio is easily measured through standard amino-acid analysis techniques (see Chapter 4; Bada, 1984).

D-Aspartic acid is introduced through racemisation into eye-lens protein in the living organism at the rate of 0.14% per year, so that a 30-year-old person has accumulated 4.2% D-aspartic acid in this particular protein. It is in age determination of recently deceased corpses (and other 'scene-of-the-crime' artefacts, for which ^{14}C-dating is inaccurate), that forensic science interest in reliable amino-acid dating is centred. For older specimens, the method is wildly unreliable: thus, Otztal Ice Man – the corpse found at Hauslabjoch, high in the Austrian Tyrol, in 1991 – was dated to 4550 ± 27 BC by radiocarbon dating, but would have a grossly inaccurate assignment of birthday on the basis of amino-acid racemisation data (Bonani *et al.*, 1994). In unrelated areas, amino-acid racemisation has given useful information on the age of art specimens (e.g. dating of oil paintings through study of the egg-protein content).

Such inferences derive from data on the kinetics of racemisation, measured in the laboratory (described in Section 4.18.2) and there is a good deal of controversy surrounding the dating method since no account is taken of the catalytic influence on racemisation rates of molecular structures that surrounded the amino-acid residue for some or all the years. It is, for example, now known that the rate of racemisation of an amino acid, when it is a residue in a protein, is strongly dependent on the nature of the adjacent amino acids in the sequence; the particular amino acid on which measurement is made might have been located in a racemisation-promoting environment for many years after the death of the organism.

1.12 Roles for amino acids in chemistry and in the life sciences

1.12.1 Amino acids in chemistry

The physiological importance of α-amino acids ensures a sustained interest in their chemistry – particularly in pharmaceutical exploration for new drugs, and for their synthesis, reactions and physical properties. As is often the case when the chemistry of a biologically important class of compounds is being vigorously developed, an increasing range of uses has been identified for α-amino acids in the wider context of stereoselective laboratory synthesis (including studies of biomimetic synthetic routes).

1.12.2 Amino acids in the life sciences

Apart from their main roles, particularly their use as building blocks for condensation into peptides and proteins, α-amino acids are used by plants, fungi and bacteria as biosynthetic building blocks. Many *alkaloids* are derived from phenylalanine and tyrosine (e.g. Figure 1.6; and *penicillins* and *cephalosporins* are biosynthesized from tripeptides, Chapter 8).

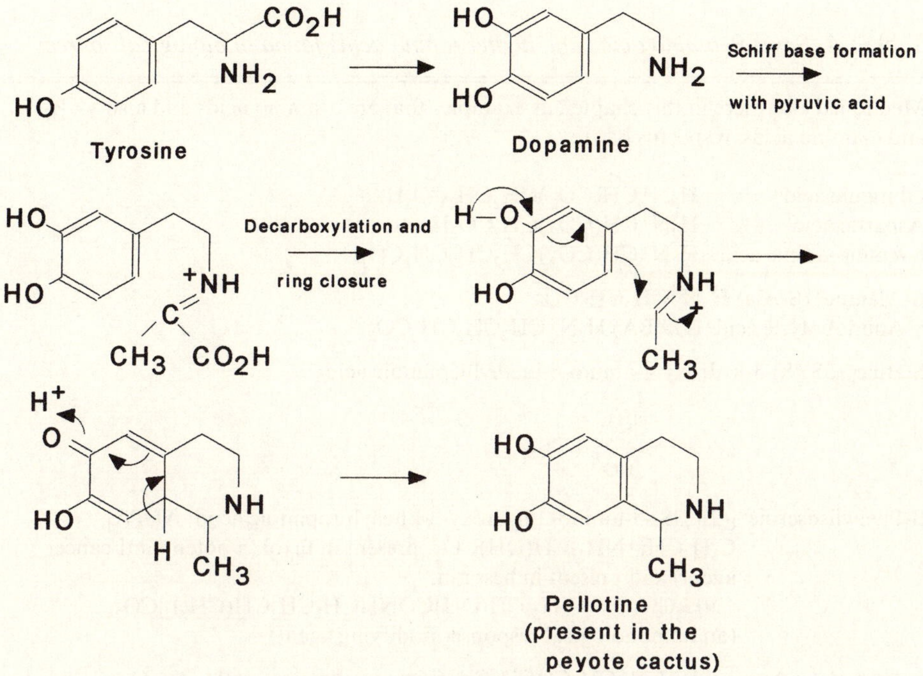

Figure 1.6. Routes from L-tyrosine to alkaloids. Alkaloid biosynthesis is often grouped into categories based on the initiating amino acid; i.e. the ornithine/cysteine route (e.g. nicotine); the phenylalanine/tyrosine/tryptophan route (e.g. the isoquinoline alkaloids, such as pellotine); etc.

1.13 β- and higher amino acids

There are relatively few examples; but there are increasing numbers of amino acids with greater separation of the amino and carboxy functions that have been found to play important biological roles (Drey, 1985; Smith, 1995). The coded amino acid, aspartic acid, could be classified either as an α- or as a β-amino acid. Glutamic acid (which can be classified either as an α-amino acid or as a γ-amino acid) is the biological source, through decarboxylation, of γ-aminobutyric acid (known as GABA; see Table 1.4), which functions as a neurotransmitter (as do glycine and L-glutamic acid and, probably, three other coded L-α-amino acids). The simple tripeptide glutathione (actually, an isopeptide; see Section 1.2 and Chapter 8) is constructed using the side-chain carboxy group rather than the α-carboxy group of glutamic acid and therefore could be said to be a peptide formed by the condensation of a γ-amino acid and two α-amino acids.

Numerous natural peptides with antibiotic activity and other intensely potent physiological actions incorporate α- and higher amino acids, as well as highly processed coded amino acids. The microcystins, which act as hepatotoxins, provide one

17

Table 1.4. *Some β-amino acids and higher amino acids found in biological sources*

Mentioned elsewhere in this chapter, as examples that are α-amino acids and also γ-, β- and δ-amino acids, respectively, are

Glutamic acid[b] $H_3N^+CH(CO_2^-)CH_2CH_2CO_2H$,
Aspartic acid[b] $H_3N^+CH(CO_2^-)CH_2CO_2H$ and
δ-Amino-adipic acid[c] $H_3N^+CH(CO_2^-)CH_2CH_2CH_2CO_2H$

β-Alanine[b] (β-Ala) $H_3N^+CH_2CH_2CO_2^-$
γ-Aminobutyric acid[a] (GABA) $H_3N^+CH_2CH_2CH_2CO_2^-$

Statine[c] (3S,4S)-3-hydroxy-4-amino-6-methylheptanoic acid

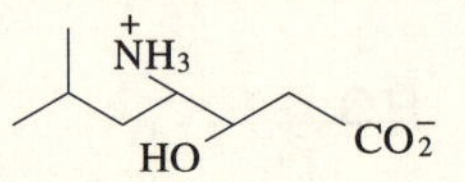

β-Phenylisoserine[c] [(2R,3S)-3-amino-2-hydroxy-3-phenylpropanoic acud; AHPA],
$\qquad$ $C_6H_5CH(^+NH_3)CH(OH)CO_2^-$, present in taxol (a potent anti-cancer agent) and present in bestatin,
$\qquad$ $^+NH_3CH(CH_2C_6H_5)CH(OH)CONHCH[CH_2CH(CH_3)_2]CO_2^-$
$\qquad$ (an immunological response-modifying agent)

δ-Aminolaevulinic acid[a] $H_3N^+CH_2COCH_2CH_2CO_2^-$ (an analogue with a C=C grouping is the active constituent of light-activated ointments for the treatment of skin cancer)

Notes:
Some of these naturally occurring amino acids are:
[a] found only in the free state and not found in peptides;
[b] found in the free state and also found in peptides; and
[c] found only in peptides and other derivatised forms.

example. They are represented by the family structure cyclo[—D-Ala—X—D-MeAsp—Z—Adda—D-Glu—Mdha—), where X and Z are various coded L-amino acids and D-MeAsp is D-erythro-β-methylaspartic acid, found in the water bloom-forming cyanobacterium *Oscillatoria agardhii*. The structure of one of these, [D-Asp³,DHb⁷]microcystin-RR (Sano and Kaya, 1995), is displayed in Chapter 3 (Figure 3.6).

Moving away from the simpler α-amino acids as constituents of peptides, the γ-amino acid (R)-carnitine $Me_3N^+CH_2CH(OH)CH_2CO_2^-$, is a rare example of a free amino-acid derivative with an important physiological role. This amino acid betaine is sometimes called 'vitamin B_T' and plays a part in the conversion of stored body fat into energy, through transport of fat molecules of high relative molecular mass to the sites of their conversion.

The (2R,3S)-phenylisoserine side-chain at position 13 of the taxane skeleton in the anti-cancer drug taxol (from the yew tree) is essential to its action.

1.14 References

Reviews providing information on all aspects of amino-acid science (Barrett, 1985; Greenstein and Winitz, 1961; Williams, 1989) and peptide chemistry (Jones, 1991) are listed at the end of the Foreword. References cited in the text of this chapter are the following.

Bada, J. L. (1984) *Methods Enzymol.*, **106**, 98.

Baldwin, J. E., Maloney, M. G. and Parsons, A. F. (1990) *Tetrahedron*, **46**, 7263.

Bonani, G., Ivy, S. D., Hajdas, I., Niklaus, T. R. and Suter, M. (1994) *Radiocarbon*, **36**, 247.

Chan, S. J. and Steiner, D. F. (1977) *Trends Biochem. Sci.*, **2**, 254.

Collingridge, G. L. and Watkins, J. C. (1994) *The NMDA Receptor*, Second Edition, Oxford University Press, Oxford.

Drey, C. N. C., in Barrett, G. C., Ed. (1985) *Chemistry and Biochemistry of the Amino Acids*, Chapman & Hall, London, p. 25.

Finkelstein, J. D. (1990) *J. Nutr. Biochem.*, **1**, 228.

Food and Drugs Administration, Washington, USA (1992) *Safety of Amino Acids used as Dietary Supplements*.

Hanby, W. S. and Rydon, H. N. (1946) *Biochem. J.*, **40**, 297.

Kothakota, S., Mason, T. L., Tirrell, D. A. and Fournier, M. J. (1995) *J. Am. Chem. Soc.*, **117**, 536.

Kyte, J. and Doolittle, R. F. (1985) *J. Mol. Biol.*, **157**, 105.

Man, E. H. and Bada, J. L. (1987) *Ann. Rev. Nutr.*, **7**, 209.

Matsusita, I., and Ozaki, S. (1993) *Peptide Chemistry; Proceedings of the 31st International Conference*, pp. 165–8 (*Chem. Abs.* **121**, 77 934).

Meldrum, B. S. (1991) *Excitatory Amino Acid Antagonists*, Blackwell, Oxford.

Noren, C. J., Anthony-Cahill, S. J., Griffiths, M. C. and Schultz, P. G. (1989) *Science*, **244**, 182.

Richmond, M. H. (1972) *Bacteriol. Rev.*, **26**, 398.

Sano, T. and Kaya, K. (1995) *Tetrahedron Lett.*, **36**, 8603.

Smith, B. (1995) *Methods of Non-α-Amino Acid Synthesis*, Dekker, New York.

Smith, M. J., Mazzola, E. P., Farrell, T. J., Sphon, J. A., Page, S. W., Ashley, D., Sirimanne, S. R., Hill, R. H. and Needham, L. L. (1991) *Tetrahedron Lett.*, **32**, 991.

Stadtman, T. C. (1996) *Ann. Rev. Biochem.*, **65**, 83.

Sutton, C. H., Baize, M. W. and Todd, L. J. (1993) *Inorg. Chem.*, **33**, 4221.

2

Conformations of amino acids and peptides

2.1 Introduction: the main conformational features of amino acids and peptides

This topic has been thoroughly developed insofar as *the conformational behaviour of amino acids and peptides in aqueous solutions* is concerned. The main driving force for conformational studies has been the pharmaceutical interest in the interactions of biologically active amino acids and peptides with tissue, particularly with cell receptors. The solid-state behaviour of amino acids and peptides, though less relevant in the pharmaceutical context, has not escaped investigation. This is because of the wider distribution and greater ease of use of X-ray crystallography equipment nowadays.

The conformational behaviour of *N*- and *C*-terminal-derivatised amino acids and peptides in *organic solvents* has also been studied, particularly by nuclear magnetic resonance and circular dichroism spectrometric techniques (in which advances in instrumentation have been very considerable; see Chapter 3).

2.2 Configurational isomerism within the peptide bond

The amide group shows restricted flexibility because its central —NH—CO— bond has some double-bond character due to resonance stabilisation [—NH—CO— ↔ —N⁺H=C(O⁻)—]. The energy barrier that this creates is insufficient to prevent rotation, but sufficient to ensure that geometrical isomers exist under normal physiological conditions of temperature and solvent, so ensuring that a particular peptide can exist in a variety of conformations, often an equilibrium mixture of several conformations, in solutions.

Planar *cis* and *trans* isomers (Figure 2.1(a)) are the most stable configurations, because the planar structure involves maximum orbital overlap. For the majority of peptides built up from α-amino acids, the amide bond adopts the *trans* geometry. α-Imino acids (notably proline but also *N*-methylamino acids), as well as α-methyl-α-

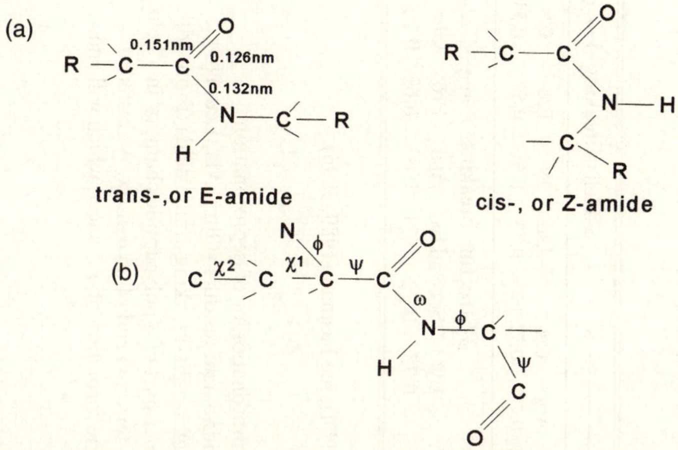

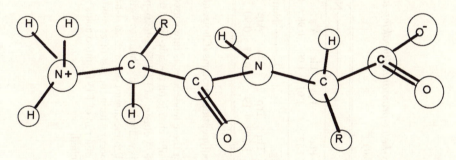

Figure 2.1. (a) Amide bonds in the *trans* and *cis* configurations. (b) Torsion angles for an amino-acid residue in a peptide.

Figure 2.2. A representative dipeptide made up from L-α-amino acids, in the extended conformation with the amide bond in the *trans* configuration.

amino acids, assist the adoption of more flexible structures by peptides when they are built into peptides, because mixtures of *cis* and *trans* configurations are more likely. *Cis*-amide bonds are rare in natural polypeptides that contain no proline residues (there are three *cis*-amide bonds in the enzyme carboxypeptidase A and one in the smaller polypeptide concanavalin A), though current re-investigations by NMR methods (Chapter 3) are revealing more distorted *trans*-amide bonds in structures established without such details in the early days of X-ray crystallography.

Regions of a peptide can exist either in a *random conformation* (i.e., the *denatured state*) or in one of a number of stereoregular conformations: the *extended conformation* (Figure 2.2), the α-*helix* (either right-handed or left-handed) and the β-*sheet* (Figure 2.3). Two of the stereoregular conformations are stabilised by intra-molecular hydrogen bonding and intermolecular hydrogen bonding accounts for

Table 2.1. *The tendency of coded amino acids for α-helix-promoting/breaking and β-structure promoting/breaking, when part of a peptide (neutral=≃0.8–1.00=no tendency either way)[a]*

←←←←←←← α-Helix-promoting (helicogenic)

Glu+	Ala	Leu	His+	Met	Gln	Trp	Val	Phe
1.53	1.45	1.34	1.24	1.20	1.17	1.14	1.14	1.12

←←——————————— Neutral ———————————→

Lys+	Ile	Asp−	Thr	Ser	Arg+	Cys
1.07	1.00	0.98	0.82	0.79	0.79	0.77

α-Helix-breaking→→

Asn	Tyr	Pro	Gly
0.73	0.61	0.59	0.53

←←←←←←← β-Structure promoting

Met	Val	Ile	Cys	Tyr	Phe	Gln	Leu	Thr	Trp
1.67	1.65	1.60	1.30	1.29	1.28	1.23	1.22	1.20	1.19

←——————— Neutral ———————→

Ala	Arg+	Gly	Asp−	Ser
0.97	0.90	0.81	0.80	0.79

β-Structure breaking→→→→→

Lys+	His+	Asn	Pro	Glu−
0.75	0.71	0.65	0.62	0.26

Notes:

[a] On the basis of a scrutiny of crystal structures of 15 polypeptides/proteins (Chou and Fasman, 1974); see Fasman (1989, 1996).

Further details:

1. An early term, 's-values' (Zimm and Bragg, 1959) for these tendencies, is rarely used now.
2. Glu+, His+, Lys+, Asp−, Arg+ indicate the presence of a charge on the side-chain as a result of protonation or of deprotonation.
3. The Chou–Fasman rules in brief (Fasman, 1985), are 'A cluster of FOUR α-helix-promoting amino-acid residues (Glu+, Ala, Leu, His+, Met, Gln, Trp, Val, or Phe) in a run of SIX amino acids will initiate an α-helix to form, until sets of α-helix-breaking amino acids (those with 's-values' less than 1.00 in this table) are encountered. Proline cannot occur in an α-helix in the inner part of a polypeptide chain, or in a helix in the C-terminal end of a polypeptide chain, but can occur within the last three residues in the *N*-terminal end of an α-helix. A cluster of THREE β-structure-forming amino-acid residues out of a run of FIVE amino acids will initiate the formation of a β-sheet, which will end when a set of four β-sheet-breaking amino acids is reached.'

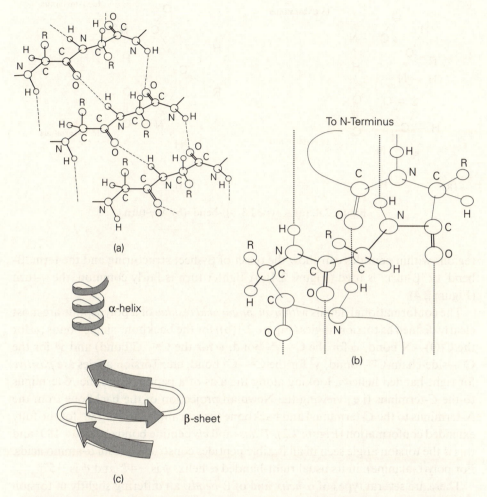

Figure 2.3. Representative peptides made up from L-α-amino acids. (a) The structural formula of a parallel β-sheet showing H-bonds. (b) The structural formula of a right-handed α-helix showing H-bonds. (c) The standard 'tape' depictions of the right-handed α-helix and antiparallel β-sheet. All amide bonds are in the *trans* configuration.

numerous physical phenomena (gel formation as gelatin solutions are cooled, which is mimicked by the behaviour of some synthetic oligopeptide solutions) and more subtle aspects of protein behaviour of physiological importance.

The α-helix is one of the best-known regular conformational features as a sub-heading within the secondary structure of polypeptides and is frequently adopted in chains of six or more *helicogenic* amino acids (see Table 2.1 for a definition of terms and examples). The β-sheet is another classic conformational structure that has been detected from the earliest days of X-ray crystallography of proteins. Local

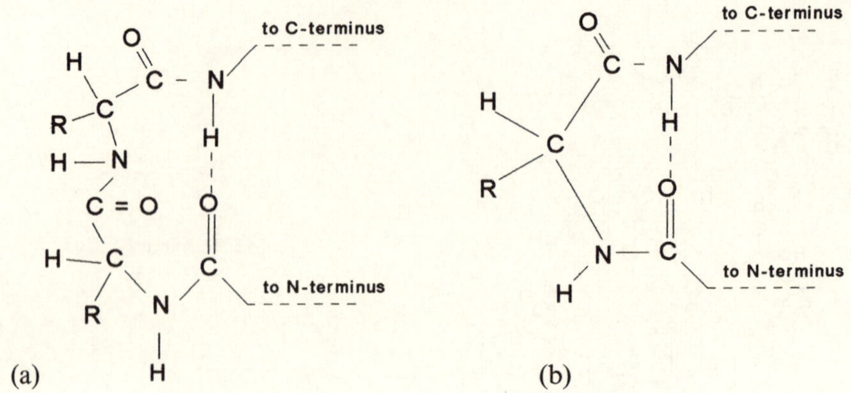

Figure 2.4. (a) A type I 3_{10}-β-bend. (b) A γ-turn.

regions within peptides can show the onset of β-sheet structuring and the term 'β-bend' or 'β-turn' is used (Figure 2.4). A tighter turn is fairly common, the γ-turn (Figure 2.4).

The conformational details *within an amino acid residue* of a polypeptide are most clearly defined as torsion angles (Figure 2.1(b)) for the backbone single bonds (ω for the C(O)—N bond, ϕ for the C^α—N bond, ψ for the C^α—C bond) and χ^1 for the C^α—side-chain-C^β—bond, χ^2 for the C^β—C^γ bond, etc. Torsion angles are *positive* for right-handed helicity, looking along the axis of a peptide from the *N*-terminus to the *C*-terminus (i.e., viewing the Newman projection of the backbone from the *N*-terminus to the *C*-terminus) and backbone torsion angles are all 180° for the fully extended conformation (Figure 2.2). *Trans*- and *cis* peptide bonds have $\omega = 180°$ and this is the torsion angle seen in all flexible peptides constructed from α-amino acids. For poly(L-alanine), in its usual right-handed α-helix, ψ is −47° and ϕ is −57°.

There are several types of α-*helix* and of β-*bends*, all differing slightly in torsion angles within the residues of the actual turn and, in the case of the α-*helix*, also determined by the pattern of intramolecular and intermolecular hydrogen bonding. Thus the type I 3_{10} β-bend includes a hydrogen bond linking three amino acid residues into a ten-membered ring; the type II variant mentioned in older literature is the same except for the helicity within the turn. There are actually *thirteen* distinct types of β-bend, differing only in dihedral angles ψ and ϕ (Hermkens *et al.*, 1994) and research into creating synthetic peptide mimetic drugs (Chapter 9) has focussed on copying the general disposition of functional groups around turns in biologically active peptides, to create analogues but often without the peptide backbone and therefore capable of reaching the target site in the living organism by virtue of surviving enzymatic destruction.

The currently interesting motif RGDS (i.e. Arg—Gly—Asp—Ser—) present in some cell-adhesion proteins that are of crucial importance in growth and other aggregation phenomena has stimulated synthetic efforts with a view to obtaining smaller molecules with similar properties. The glycine and aspartic acid residues form part of a β-turn, when the RGDS sequence is part of a small peptide (see also Section 2.4).

The α-helix can enclose more than the average 3.4 amino-acid residues per turn that is its standard feature and can exist in left-handed helicity as well as in its most common right-handed form (the *right-handed* form is normally induced by the L-configured chiral centres).

Imino-acid residues modify these torsion angles considerably; replacement of the amide proton, for example by methylation, induces conformational changes, since *cis–trans* isomerisation is more easily brought about (because there is a lower energy barrier). However, hydrogen-bonding ability and, of course, other types of close-packing interactions, are substantially modified at α-imino-acid residues.

The conformations of natural polypeptides have arisen as a consequence of the sequences of amino acids that they contain. In some cases, their biological function does not depend directly on the presence of particular amino acids (e.g., collagen and other connective tissue) whereas, in other cases, certain functional side-chains are mandatory at particular sites in the molecule in order for physiological activity to be shown (enzymes). The correct three-dimensional disposition of these side-chains is obtained through conformational features in their vicinity within the poly-peptide and often through features some distance away (Chapter 8).

The conservatism implied in protein structure is revealed with the effect of *N*-methylation at one amino-acid residue in insulin (conversion of an α-amino-acid residue into an α-imino-acid residue). Residue 2 in the A-chain is L-isoleucine (the structure is given in Figure 1.4) and replacement of this by its *N*-methyl analogue, or similar replacement of residue 3 (L-valine), leads to distorted helical sequences from residues 2–8, as shown by circular dichroism (CD) measurements, which leads in turn to the change-over from the natural quaternary structure (dimeric for bovine zinc insulin) to the monomeric form. It is not surprising, bearing in mind the sub-stantial conformational changes accompanying apparently insignificant structural changes, that these analogues show only about 12% of the potency of the natural hormone, according to radioimmunoassay evidence (Chapter 4).

The effect of a *trans→cis* change at just one position in a biologically active oligopeptide can drastically change the properties of that peptide. The antibiotic Gramicidin S, with its alanyl residue replaced by α-amino-isobutyric acid (Ala→Aib, causing a *trans→cis* configurational change at the amide bond at that location according to CD measurements) is inactive against Gram-positive bacte-ria. Insertion of Aib into a peptide generally restricts the local conformation to a left-handed or right-handed 3_{10} β-bend or α-helix, the outcome depending on pre-cisely what other residues are nearby.

Structural changes of this nature need, however, not have an effect if the changes are remote from the part of the peptide that responds to receptors; thus bradykinin

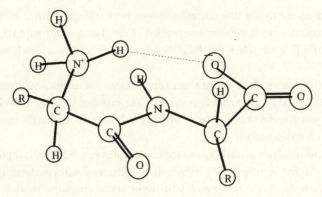

Figure 2.5. The conformation of a D,L-dipeptide in water.

(Section 4.11) with the Pro→Aib substitution at position 7 retains high biological activity, probably because bradykinin itself has a *cis*-amide grouping at the *N*-terminal side of this proline residue (Cann *et al.*, 1987). The conformational nature of natural bradykinin is now well understood and, under normal physiological conditions, an individual molecule spends its time interconverting between two types of conformation, one disordered and the other partially ordered, with a right-handed 3_{10} β-bend in the *N*-terminal region and a β-turn towards the *C*-terminus (Cann *et al.*, 1987).

2.3 Dipeptides

The flexibility of acyclic structures such as peptides has been established convincingly in the broad picture that has emerged from conformational studies. The solution conformation of a dipeptide is sensitive towards (a) the nature of the solvent, (b) the concentration of the solution, (c) the temperature of the solution and (d) the presence and nature of other solutes. The sensitivity is much greater for dipeptides than it is for other acyclic solute molecules, owing to the proximity of the terminal amino and carboxy groups. A further controlling factor is configuration; there are four possible stereoisomers for a representative dipeptide, namely two diastereoisomers and their enantiomers. An L,L-dipeptide (Figure 2.2) adopts a quite different solution conformation from that of its D,L-diastereoisomer (Figure 2.5).

2.4 Cyclic oligopeptides

Beyond the relatively rigid cyclic dipeptides (dioxopiperazines; see Chapter 6), macrocyclic peptides show considerable flexibility though relatively less flexibility than that of their acyclic analogues. The biological importance of cyclic peptides has been attested by numerous examples (antibiotics such as Gramicidin S and immunosuppressant agents such as cyclosporin; Figure 1.4).

Many attempts to restrict conformational freedom of amino-acid and peptide analogues synthesised for pharmaceutical studies have been described in reports of current drug research, one such approach being the synthesis of cyclic analogues of biologically active short peptides. For example, the importance of the cell-adhesion sequence RGDS (i.e., —Arg—Gly—Asp—Ser—; see also Section 2.2) has stimulated research in which cyclic peptides enclosing it in such a manner as to constrain its conformation have been synthesised for biological testing (the fact that there are naturally occurring cyclic peptides possessing biological activity provided the inspiration for this research).

2.5 Acyclic oligopeptides

As the distance of separation of the terminal functional groups increases, the sensitivity of the conformation towards solution parameters decreases. Attractions between side-chains become more dominant in determining conformations of longer peptides; these interacting side-chains do not necessarily have to be close together if they are polar, but may be 'bridged' by water molecules.

Oligopeptides are generally water-soluble, the hydrophilic amide backbone groupings contributing substantially to this property, and intramolecular attractions within short oligopeptides are practically nonexistent in water. *N*- and *C*-derivatised dipeptides are insoluble in water, but derivatised peptides longer than three or four amino-acid residues are usually appreciably soluble in water, owing to their greater number of backbone amide groups.

2.6 Longer oligopeptides: primary, secondary and tertiary structure

As the chain length increases further, other factors come into play to determine the overall conformation of the peptide and the electrically charged groups at the ends of the chain are less important in this respect. Interactions, repulsive and attractive, between side-chains are dominant and the *primary structure* (the sequence of the peptide and the stereochemistry at each chiral centre) determines the run of the peptide chain through the molecule (the *secondary structure*) and the overall shape of a single polypeptide chain (globular, extended, etc.; the *tertiary structure*) of the molecule.

The term '*domain*' is applied to describe regions within a protein molecule. Particular associations can arise between distinct secondary structures within the tertiary structure (an α-helix in one part of the polypeptide molecule can pack with a particular region elsewhere in the sequence; a β-sheet can be stabilised by a nearby structural feature from a distinct part of the polypeptide sequence). These domains

27

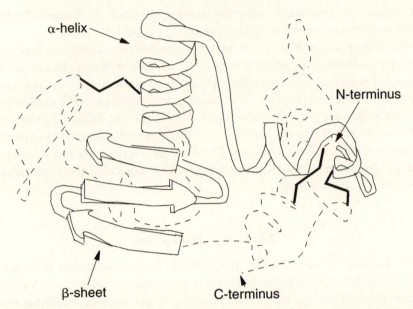

α-helix

N-terminus

β-sheet

C-terminus

Figure 2.6. Egg lysozyme (a globular protein), illustrating the method of representing the conformation of a polypeptide, in which the run of the polypeptide through the molecule is made clear by depicting it as a ribbon. There are three disulphide bridges in this enzyme. Parts of the molecule are shown in greater detail, to reveal that a globular protein is typically made up of a collection of the various conformational features (α-helix and β-sheet – with β-bends determining the number and length of each stretch of extended conformation – and random regions).

are usually dominant stabilising features, characteristically leading to the functional conformation of a protein (Figure 2.6).

2.7 Polypeptides and proteins: quaternary structure and aggregation

The association of two or more identical polypeptide molecules to form an overall globular protein is often found to occur for enzymes and other proteins, the resulting overall entity representing their quaternary structure. This association behaviour can be demonstrated using synthetic peptides. Spectacular examples have been worked out by reasoning out the structural requirements needed, on the basis of the behaviour of individual amino acids – thus, derivatised *amphiphilic oligopeptides* (peptides with alternating hydrophilic and hydrophobic residues, e.g. the *N*-acetyl hexadecapeptide amide Ac—EAALEAALELAAELAA—NH$_2$) form aggregates of α-helices in aqueous solution at pH 7 (0.5 mg ml^{-1}) which change to aggregated β-sheets when the pH of the solution is lowered to 4 (Mutter *et al.*, 1991). The evidence for this association behaviour is easily obtained through circular dichroism spectroscopy (Chapter 3). The side-chains of the glutamic acid residues (E) and of the lysine residues (L) are on opposite sides of the peptide backbone in its extended conformation (Figure 2.2) and the helicogenic nature of the constituent amino acids at pH 7 causes the formation of α-helices that present a hydrophilic face on one side of the helix and their hydrophobic face on the other (Figure 2.7).

2.8 Examples of conformational behaviour; ordered and disordered states and transitions between them

There are direct consequences of the amino-acid make-up and sequence of amino-acid residues within a peptide (its primary structure) on its overall conformational behaviour. In spite of the regularity of the backbone of the molecule, a wide variety of conformations is adopted, indicating the role of the side-chains in determining the overall conformation of the molecule.

2.8.1 The main categories of polypeptide conformation

Sequence-determined features are summarised in the following sub-sections.

2.8.1.1 One extreme situation

Simple aliphatic hydrocarbon side-chains tend to lead to adoption of a regular conformation. The side-chains are *hydrophobic* and repel each other but repel water to a much greater extent – the molecule organises itself so that *amide groups* undertake stabilising hydrogen bonding with each other, leading to the α-helix, or (if the side-chains are small enough) the β-sheet is favoured.

2.8.1.2 The other extreme situation

Polar side-chains tend to lead to adoption of an irregular (random; denatured) conformation. The side-chains are *hydrophilic* and the overall molecule gains stabilisation through interactions between the side-chains and a polar solvent (such as water); furthermore, *amide groups* do not easily hydrogen bond with each other.

2.8.1.3 The general case

For polypeptides with irregular sequences, as is the usual case for enzymes and other water-soluble globular proteins, the conformational situation ranges from totally random to a mixed situation with regular stretches intermingled with random (irregular) lengths of the backbone.

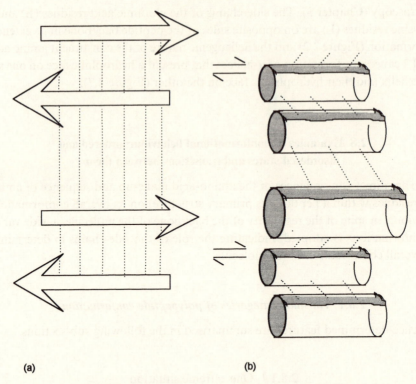

(a)   (b)

Figure 2.7. Aggregated conformations of an amphiphilic oligopeptide. (a) Aggregated antiparallel β-sheets at pH 4. (b) α-helices aggregating at pH 7.

2.9 Conformational transitions for peptides

For the general case, conformational transitions can be very easily brought about. The changes occur even more easily for certain particular cases such as poly(α-amino acid)s in which the side-chain has a functional group whose state of ionisation can be altered (Figure 2.7).

The ordered–disordered transition (denaturation) is common behaviour for polypeptides and proteins which can often be brought about thermally or by changes to the solvent, such as changes of ionic strength or of pH. It is often easily reversible; thus, the poly(L-glutamic acid) side-chain is negatively charged in solutions at pH 7, whereas the poly(L-lysine) side-chain is positively charged at this pH; both these poly(L-amino acid)s are in the random conformation in aqueous solutions, until, by lowering the pH in the former case and raising the pH in the latter case, ordering accompanies the neutralisation of the charges on the polymers and they adopt the right-handed α-helix conformation.

2.10 References

Cann, J. R., London, R. E., Unkefer, C. K., Vavrek, V. J. and Stewart, J. M. (1987) *Int. J. Pept. Protein Res.*, **29**, 486.

Chou, P. Y. and Fasman, G. D. (1974) *Biochemistry*, **13**, 222.

Fasman, G. D. (1985) *J. Biosci.*, **8**, 15.

Fasman, G. D. (1989) *Prediction of Protein Structures and the Principles of Protein Conformation*, Plenum, New York.

Fasman, G. D. (Ed.) (1996) *Circular Dichroism in Conformational Analysis of Biomolecules*, Plenum, New York.

Hermkens, P. H. H., van Dinther, T. G., Joukema, C. W., Wagenaars, G. N. and Ottenheijm, H. C. J., (1995) *Tetrahedron Lett.*, **35**, 9271.

Mutter, M., Gassmann, R., Buttkus, U. and Altman, K.-H. (1991) *Angew. Chem. Int. Ed.*, **30**, 1514.

Zimm, B. H., and Bragg, J. K. (1959) *J. Chem. Phys.*, **31**, 526.

3

Physicochemical properties of amino acids and peptides

3.1 Acid–base properties

The physicochemical properties of amino acids depend on (a) the presence of groups (e.g. amino, carboxy, thiol, phenolic hydroxy, guanidino and imidazole) that can be titrated in the pH range 0–14, (b) the presence or absence of hydrophobic groups (e.g. alkyl, aryl and indole) and (c) the presence or absence of neutral hydrophilic groups (e.g. aliphatic hydroxy and side-chain amide groups). The properties of peptides also depend on the same factors, but it must be remembered that, in a linear peptide containing n amino-acid residues, all but one α-amino group and one α-carboxy group are incorporated into neutral peptide and amide bonds. In a cyclic peptide, there are no free α-amino or α-carboxy groups. Moreover, some peptides contain groups such as carbohydrate, phosphate ester, lipids and porphyrins that further modify physical properties.

A simple amino acid exists in neutral aqueous solution as a dipolar ion (*Zwitterion*), $NH_3^+CHRCO_2^-$. The carboxy group has a pK_a value of approximately 2.3 and is about 300 times stronger an acid than is acetic acid due to the electrostatic effect of the $-NH_3^+$ group. The pK_a of the $-NH_3^+$ group in an α-amino acid ester is about 7.7. This should be compared with the pK_a (about 10.6) for a simple aliphatic primary amine. The large difference results from the powerful electron-attracting effect of the $-COOR$ group. The $-COO^-$ group in the dipolar ion of an amino acid cannot have such a large effect since it already has a negative charge. Nevertheless, there is still an appreciable electron-withdrawing effect and the pK_a of the $-NH_3^+$ in an α-amino acid is about 9.7. In other words, the $-NH_3^+$ group in an α-amino group is about eight times stronger as an acid than is the $-NH_3^+$ group in a salt of an aliphatic primary amine. In peptides and proteins, the peptide bond nearest to the *N*-terminal $-NH_3^+$ group is more strongly electron-attracting than is the $-COO^-$ group in an α-amino acid, so the terminal $-NH_3^+$ group is more acidic

(pK_a 7.4–7.9) than is that in an α-amino acid. Likewise, the terminal α-carboxy group in a peptide or protein is also influenced by the proximal peptide bond, but less so than it is by the —NH_3^+ group in an α-amino acid. The pK_a is 3.1–3.8, which is intermediate in acid strength between the carboxy groups of acetic acid and glycine. In contrast, the ε-NH_3^+ group in the side-chain of lysine and the β- and γ-carboxy groups of aspartic and glutamic acids respectively are more or less insulated from these electronic effects by intervening saturated carbon atoms. The pK_a of the ε-NH_3^+ of a lysyl residue is about 10.2 and the pK_a values of the ω-CO_2H groups in aspartic and glutamic acids are 3.9 and 4.3 respectively.

There are several other titratable groups in peptides and proteins. The imidazole group in the side-chain of histidine can be protonated and the imidazolium moiety has a pK_a of 6.0–7.4 depending on the proximity of other groups (see below). The thiol group in the side-chain of cysteinyl residues is weakly acidic (pK_a 8.5–10.4) and the guanidinium group of arginine is so weakly acidic ($pK_a > 12$) that its titration with alkali usually denatures a protein.

The interpretation of titration curves of peptides and proteins can be quite tricky. In addition to the number of groups that may be involved, their pK_a values can be perturbed by several factors. For example, when charged groups are in close proximity and when salts are present, pK_a values are influenced by electrostatic effects. Titration thus gives apparent pK_a values and the intrinsic values have to be computed by applying a correction factor based on the Debye–Hückel theory:

$$pK = pK_{intrinsic} - 0.86wZ,$$

where Z is the net charge (i.e. the algebraic difference between the numbers of positive and negative charges) and w is given by the equation

$$w = \frac{Ne^2}{2DRT}\left(\frac{1}{b} - \frac{\kappa}{1 + \kappa\alpha}\right) = 3.57\left(\frac{1}{b} - \frac{\kappa}{1 + \kappa\alpha}\right)$$

in water at 25 °C, where N is Avogadro's number, e is the electronic charge, D is the dielectric constant of the medium, R is the gas constant, T is the absolute temperature, b is the radius of the charged peptide or protein (in ångström units) assuming that it is spherical, α is the distance of closest approach of a small ion to the peptide or protein sphere ($\simeq b + 2.5$Å) and κ is the Debye–Hückel parameter defined by the equation

$$\kappa = \left(\frac{4\pi Ne^2 \sum_i c_i z_i^2}{DRT}\right)^{1/2}$$

where c_i and Z_i are the concentration and charge of the ith ionic species. At best, this correction is only approximate since it assumes that the peptide or protein is spherical with the charges uniformly distributed over the surface. In addition, the effective dielectric constant in the vicinity of the molecule is likely to be much lower than D, the bulk dielectric constant of the solvent.

In addition to electrostatic effects, the presence of adjacent hydrophobic groups or hydrogen bonds can perturb the pK_a of titratable groups. If a titratable group is adjacent to hydrophobic groups, it will behave as if it were dissolved in an apolar solvent of low dielectric constant. Dissociation of a carboxy group into the carboxylate anion and a proton occurs much less readily in a solvent of low dielectric constant than it does in water, since two charged ions are produced from an uncharged group. Consequently, the pK_a is increased. In contrast, the dissociation of a proton from a $-NH_3^+$ results in no net change in the number of charged particles and, consequently, the pK_a value of such a group is relatively insensitive to a change of dielectric constant.

When hydrogen-bonding involves titratable groups, the pK_a may be increased or decreased according to circumstances. If the acidic form of an acid is acting as a donor, removal of the proton will be inhibited and the pK_a will be increased. Conversely, if the basic form of a conjugate acid–base system is acting as an acceptor, addition of a proton to the base will be inhibited and the pK_a will be lowered.

Before the advent of NMR spectroscopic methods for determining the secondary structure of proteins, detailed analysis of the acid–base properties of peptides and proteins was one of the few techniques available for acquisition of such knowledge. Nowadays, investigation of acid–base properties tends to be limited to the study of groups that are believed to be involved in the manifestation of biochemical and biological properties. Before ending this section, it is worth mentioning the special case of the phenolic hydroxy groups of tyrosyl groups, the pK_a values of which can be determined uniquely by spectrophotometric titration.

3.2 Metal-binding properties of amino acids and peptides

Amino acids are bidentate ligands for several transition metal ions. Consequently, Cu^{2+} ions can be used to form complexes as described in Section 7.5 in order to block functional groups selectively. Useful intermediates such as ε-Z- and ε-Boc—Lys—OH can thus be readily prepared. The —CONH— groups in peptides also complex with Cu^{2+} ions and this forms the basis of the classical biuret reaction for peptides and proteins. Some of the side-chains in peptides and proteins, especially those of Glu and His, are good ligands for transition metal ions. Several enzymes such as collagenase, stromelysin, carboxypeptidase and carbonic anhydrase contain Zn^{2+} ions that are complexed with functional groups in the side-chains of amino acids such as His and Glu. A distinction should be made between cases such as those cited above, in which metal ions complex with the amino acids in a peptide chain, and proteins such as haemoglobins and myoglobins, in which Fe^{2+} is complexed in a porphyrin system which is itself non-covalently bound to a polypeptide chain. It is important to note that, upon binding oxygen in the lung, the iron in haemoglobin remains in the Fe^{2+} state. If the iron is converted into the Fe^{3+} state, as it can be when exposed to some oxidising agents, the resulting methaemoglobin does not function as an oxygen transporter. This condition is known as methaemoglobinaemia. Many mutant forms of haemoglobin have been discovered and some of these have a strong tendency to become oxidised to the non-functional Fe^{3+} form. The ability of transition metal ions to complex with a range of groups found in peptides and proteins probably accounts in large measure for the toxicity of these cations.

Special mention must be made of the thiol group in the side-chain of cysteine, which has a high affinity for Ag^+ and Hg^{2+} ions. It will be seen later that Hg^{2+} ions are used to remove Acm protecting groups from cysteine side-chains (Section 7.5.2). The ease with which this type of reaction occurs is a consequence of this high affinity of Hg^{2+} ions for divalent sulphur.

3.3 An introduction to the routine aspects and the specialised aspects of the spectra of amino acids and peptides

Interpretation of spectrometric features of amino acids and peptides and their derivatives has its routine aspects, but the spectra of these compounds also incorporate unique detail that provides specific information about the behaviour of amino acids and peptides in solution. These methods reveal the ways in which groupings within a peptide in solution relate to each other and these details are of major importance in determining the physical and physiological properties of amino acids and peptides. Conformational studies (Chapter 2) and structure determinations for peptides have been the priority targets for investigations by physicochemical methods.

The routine category, the use of spectrometry in support of synthetic and investigative studies (verification of the course of a synthesis or of reactions of amino acids and peptides, for example) is a topic that has been covered by numerous textbooks. A thorough coverage of this topic is therefore not provided here and the reader is assumed to know the background relating to the uses of the techniques (and the reader is also assumed to be prepared to consult standard texts for any more detailed explanation that may be needed).

Mass spectrometry has particular applications in structure determination (see Chapter 4, Part 2) and is a spectroscopic technique that is usually available in modern chemistry laboratories, together, of course, with infrared (IR), ultraviolet (UV) and nuclear magnetic resonance (NMR) spectrometry. X-ray crystallographic facilities are less commonly accessible.

NMR spectroscopy can be highly informative, not only at the routine level, but also for the specialised detail it can provide when it comes to discovering the ways in which groupings relate to each other within a flexible molecule such as a peptide in solution. This even applies to the behaviour of amino acids and peptides in the living cell, in specialist applications that are currently undergoing development side-by-side with the well-known medical uses for NMR. The more sophisticated Fourier-transform instruments that are needed for the provision of solution behaviour information for oligopeptides and polypeptides are gradually becoming routinely available.

Circular dichroism (CD) of free amino acids and peptides and their derivatives provides valuable information on conformational behaviour and can define absolute configurations. The *UV fluorescence* behaviour of particular derivatives can also highlight positions of groups in molecules, in relation to the positions of other groups, providing decisive structural information. Although fluorescence spectrometry can be carried out in most research laboratories, CD spectrometers are less widely available.

As with any other area of investigative organic chemistry, the complementary nature of all these techniques can be exploited in peptide studies, giving the broadest possible range of information in modern problem-solving studies.

3.4 Infrared (IR) spectrometry

The main specialised information that flows from IR studies of amino acids and peptides is obtained from their derivatives. The insolubility of the un-derivatised compounds in solvents that are routinely used for IR spectrometry (CCl_4, $CHCl_3$, CS_2, etc.) is a major problem and the solubility of free amino acids (L-aspartic acid, 0.005 g ml^{-1}; L-tyrosine, 0.0005 g ml^{-1} at 25 °C) and of some short peptides in water is often low. The variation in the solubility of amino acids in water is considerable (L-proline, 1.62 g ml^{-1}). The solubility of short peptides in water is sequence-dependent, but longer oligopeptides are generally more soluble in water and less soluble in organic solvents, as a consequence of the increasing proportion of hydrophilic amide groups in relation to side-chains (which are often hydrophobic; see Table 1.1, Chapter 1).

The feature of IR spectra for solutions that is useful in the peptide area concerns the hydrogen-bonding property of the amide group. The characteristic carbonyl stretching frequencies of the peptide bond depend on conformation, so IR spectra therefore give some information on conformation. IR spectra can also be interpreted to detect conformational changes that occur when solution parameters are altered; such solution changes (changes of solvent polarity, ionic strength, etc.) can be deliberately designed either to disrupt or to augment hydrogen-bonding interactions and these changes lead to differing stretching frequencies of the amide group. An example of the use of IR to establish the conformational behaviour of a simple dipeptide in solution is shown in Chapter 2. A rough estimate of changes in the (α-helix plus random)/(β-sheet) ratio for human serum albumin from IR spectra indicates that increasing amounts of the protein adopt the β-sheet conformation as thermal denaturing occurs (Palm, 1970).

Uncertainty in such interpretations of IR spectra is more likely with longer pep-

Table 3.1. *Spectroscopic parameters*

$v=3360-3260$ for the N—H stretching frequency of the —CO—NH— grouping
$v=1250, 1550$ cm^{-1} for *trans*-amides
$v=1350, 1500$ cm^{-1} for *cis*-amides (the amide II band at 1500–1575 cm^{-1} is absent for
    N-alkyl amides —CO—NR—)
$v=620, 1650$ cm^{-1} for the hydrogen-bonded amide bond of the α-helix conformation
$v=700, 1630$ (strong) and 1690 (weak) for the β-sheet conformation
$v=650, 1660$ cm^{-1} for the peptide bond in random (disordered) conformations

---

tides, owing to the larger number of amide groupings in longer peptides that lead to overlapping features in IR spectra that are impossible to resolve and therefore to relate to individual amide groups (Table 3.1). In these more complex cases, pooling of data from several spectroscopic techniques and taking other physical measurements into account is more effective for solving conformational problems.

3.5 General aspects of ultraviolet (UV) spectrometry, circular dichroism (CD) and UV fluorescence spectrometry

These techniques are intimately related, since their spectral features originate in the same physical event, namely the absorption within locations (chromophores) of the amino acid or peptide of light from the ultraviolet wavelength region ($\lambda = 200-400$ nm) and from longer (visible) wavelengths for coloured compounds.

The chromophores that respond to electronic excitation which are common to amino acids and peptides are the amino, carboxy and amide groups. All these show almost no absorption, i.e. they are nearly transparent, showing only small extinction coefficients within the UV range (200–400 nm): —NH$_2$ (as in ammonia, NH$_3$, $\lambda_{max} = 194$ nm; but in primary amines, such as in methylamine, CH$_3$NH$_2$, $\lambda_{max} = 214$ nm); —CO$_2$H (as in acetic acid, CH$_3$CO$_2$H), $\lambda_{max} = 204$ nm; —CONH— (as in acetamide, CH$_3$CONH$_2$), $\lambda_{max} = 205$ nm. Insofar as the side-chains are concerned, the aliphatic examples are also transparent (—CH$_3$, λ_{max} below 185 nm; —OH, (as in methanol, CH$_3$OH), $\lambda_{max} = 182$ nm) but the side-chains of several coded amino acids contain chromophores that absorb at wavelengths longer than 200 nm with moderate-to-high intensity (Table 3.2).

Most of the uses of UV spectrometry in the field of amino acids, peptides and proteins are entirely routine. The measurements and interpretations are simple and can be useful for determining solution concentrations of proteins, expressed in mol l^{-1}, on the basis of knowledge of the overall amino-acid composition of the particular protein (Chapter 4). They depend on calculations involving the determination of the absorbance at λ_{max} near 280 nm, using the value of the molar absorptivity (ε) for constituent non-transparent amino acids in the calculation.

Table 3.2. *UV absorption features of coded amino acids that can be exploited in quantitative analysis (data for aqueous solutions at pH 6)*

Phenylalanine	λ_{max}=208 nm ε= 8000
	λ_{max}=260 nm ε= 150
Tyrosine	λ_{max}=225 nm ε= 8000
	λ_{max}=272 nm ε= 1200
Tryptophan	λ_{max}=218 nm ε=35000
	λ_{max}=281 nm ε= 5500

Therefore, simple quantitative analysis based on UV measurements is possible for many proteins, especially those containing aromatic chromophores. Analysis protocols may exploit other physical measurements, such as the titration of the phenolic hydroxy group to determine the number of tyrosine residues present in a peptide or protein, by following changes in UV spectra of a protein as a function of pH.

Modern UV spectrometers may offer the option of first and second-derivative UV spectroscopy and measurements in the 240–320 nm range using these techniques can be interpreted to determine the ratios of the various aromatic residues in peptides. Thus, tyrosine can be differentiated from tryptophan through first-derivative spectroscopy and phenylalanine can be differentiated from tyrosine and tryptophan through second-derivative spectroscopy (Miclo *et al.*, 1995).

The determination of more subtle structural features depends upon the use of the other electronic absorption techniques, CD and fluorescence in particular, often supplemented by, or supplementing, NMR and other data. Tyrosine and tryptophan are pre-eminent in fluorescence and phosphorescence studies, since quantum yields and fluorescence decay kinetics are sensitive to the environment of the side-chains when these amino acids are condensed into a polypeptide. For example, the pK value of the phenolic group in tyrosine may be changed considerably by nearby functional groups (particularly carbonyl-containing groups) within a polypeptide or protein and fluorescence measurements (with $\lambda_{emission}$ = 315 nm and $\lambda_{excitation}$ = 275 nm from UV lamp) as a function of pH can be used to extract knowledge of this sort (Lakowicz, 1992). Some analytical applications based on spectrofluorimetry after derivatisation of amino acids and peptides with fluorescent groupings ('fluorophores') are covered later (Section 4.5.1).

3.6 Circular dichroism

Chromophores in chiral environments generate circular dichroism (CD) as a consequence of the absorption of light. This CD is portrayed as a spectrum by the CD

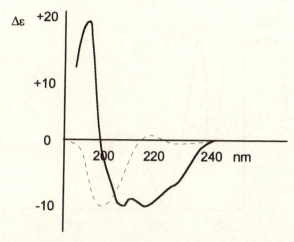

Figure 3.1. CD spectra of poly(L-glutamic acid), (——) in aqueous solution at pH 4.3
(α-helix) and (- - -) in aqueous solution at pH 11 (random conformation).

spectrometer, an instrument that measures the intensity of absorption of left-circularly polarised light relative to that of right-circularly polarised light over a continuous range of wavelengths (Figure 3.1).

The simplest CD spectrometers display the main features for the side-chains of coded aromatic α-amino acids, for example, since these features are within the easily accessible UV wavelength range. However, the additional CD data obtained for peptides and proteins by penetration to shorter wavelengths ($\lambda < 200$ nm) calls for more sophisticated instrumentation and interpretations described later in this chapter would not have been possible without this penetration.

The CD feature produced by an electronic transition within a chromophore is a simple Gaussian peak centred on the λ_{max} seen in the UV spectrum for the chromophore and the UV spectrum for such a transition within a chromophore is a simple smooth curve; conversely, the fine structure seen in the UV spectrum (e.g. for aromatic chromophores) is also superimposed on the smooth CD curve when more complex absorption features arise (numerous electronic transitions of similar energy, therefore generating absorption peaks appearing at similar wavelengths; Figure 3.2).

The amine, amide and carboxy chromophores that are common to the general family of amino acids, peptides and proteins show absorption features in the short-wavelength part of the ultraviolet range; to establish their associated CD features requires more sophisticated spectrometers. Much of the detailed conformational information gained from CD studies depends on data from this wavelength region.

The phenomenon of differences between the absorption of left- and of right-circularly polarised light is not restricted to the visible and UV wavelength regions, so infrared and Raman CD are likely to yield even more sophisticated information

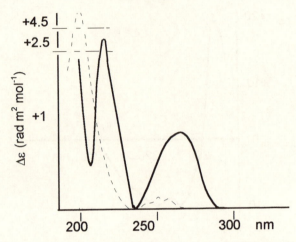

Figure 3.2. CD spectra of L-phenylalanine (---) and of L-tyrosine (——) in 0.1 M hydrochloric acid.

in the future, such as the rotational flexibility of the amino-acid side-chains. However, Raman CD instrumentation is still at the prototype stage and data are difficult to interpret even for methyl side-chains (i.e. for L-alanine); more complex cases call for understanding of the underlying principles that is not yet well-developed.

CD spectra carry much more information than do UV spectra; the intensity of the CD absorption is dependent upon the spatial relationship between the chromophore and groupings at the chiral centre and therefore there is no chromophore–intensity-of-absorption relationship such as that which exists for UV spectra (i.e. the Bouguer–Beer–Lambert law does not apply to CD spectra). Also, the sign of the CD feature can be positive or negative, unlike the isotropic absorption (i.e. the UV spectrum), which has no sign.

The CD spectrum can be interpreted in terms of absolute configuration; the sign of a particular CD feature corresponds to a particular absolute configuration of the solute, for the chiral centre nearest the chromophore responsible for that CD feature. Information on conformation (based on the sign and specific details of an overall CD spectrum for a compound of known absolute configuration) can be obtained for amino acids and peptides. An example given in Chapter 2 (Section 2.7) illustrates a simple example of this type of result.

Polypeptides in *a random conformation* show strong CD features only at short wavelengths, but characteristically enhanced CD features are observed at longer wavelengths if a molecule adopts a regular conformation and it contains a chromophore that is repeated regularly and spatially uniformly throughout the molecule (as is the case for ordered peptides; the α-*helix* and the β-*sheet structures*, Figure 3.3 and Table 3.3; see also Figures 2.2–2.5).

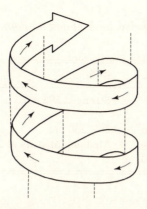

Figure 3.3. The arrows within the tape show the direction of the transition moment in each amide chromophore (α-helix; illustrative only). Coupling of transition moments between regularly arranged chromophores enhances the CD intensity.

The proportions of these three conformations that exist within a complex polypeptide or protein can therefore be determined with fair accuracy (using the Chou–Fasman rules; Chou and Fasman, 1974; see also Chapter 2). Results such as 'this protein in aqueous solution is 35% α-helical and 10% β-structured' that appear in the research literature are based on interpretation of CD spectra (Woody, 1995). A search for the particular stretches of these conformations within the polypeptide is then attempted, through deductions based on the amino-acid sequence, taking account of the locations of the hydrophilic and hydrophobic side-chains, to define the three-dimensional structure of a polypeptide.

Such conclusions have been confirmed, in some cases, by X-ray crystallographic and NMR structure determinations. Although the CD conclusions, like those derived from IR and NMR studies, are *for solution conformations*, and the X-ray crystal structure must relate to *the solid state*, in fact X-ray measurements with proteins are usually carried out on a fragile crystal in which the molecule is bathed in solvent. Typically, a crystal prepared for X-ray work will contain 50% water by weight, so the 'crystal structure' is effectively that of a molecule encased in, and thoroughly penetrated by, water. The aqueous solution conformation of the polypeptide is essentially the same as the conformation seen in its X-ray crystal structure. A representative example is egg lysozyme (Figure 2.6).

3.7 Nuclear magnetic resonance (NMR) spectroscopy

A considerable level of instrumental sophistication has to be reached in order to map out all the atoms in a complex molecule through NMR spectroscopy. However, the current literature includes examples of this topic encroaching on the role that X-ray crystallographic analysis has played for many years in providing information on

Table 3.3. *CD features for α-helix, β-sheet and random (i.e. unordered) conformations*

	Unordered	β-Sheet	α-Helix
Positive CD maximum	Weak at 218 nm	Strong at 195 nm	Very strong at 191 nm
Zero CD	211, 234, 250 nm	207, 250 nm	202, 250 nm
Negative CD maximum	Strong at 197 nm Very weak at 240 nm	Medium intensity at 217 nm	Strong at 208, 222 nm

the structures of proteins. This specialised use of NMR is the province of relatively few laboratories, but interpretation of routine Fourier-transform NMR spectra can provide considerable insights into amino-acid and peptide structure.

The simplest modern ^{1}H NMR spectrometer can provide spectra that are suitable for giving evidence for the presence or absence of some functional groups through chemical shift data and also can provide evidence, through coupling constant data, of conformational relationships (i.e. torsion angles; see Chapter 2) in the chains of carbon atoms of the $C^\alpha H$—$C^\beta H_2$— etc. sequence in the side-chain of an amino-acid residue. Particular NH protons in an oligopeptide may be shielded from hydrogen bonding to aqueous solvent and these may be identified by ^{1}H NMR spectroscopy in non-hydroxylic solvents, through ^{2}H–^{1}H-exchange studies with ^2H_2O and by measuring the temperature coefficients for NH proton resonances.

However, this information lacks reliability as soon as peptides longer than two or so residues are studied, because of the overlapping of peaks in spectra from low-magnetic-field-strength ^{1}H NMR spectrometers. Also, the limitations on solvents that are suitable for NMR studies are considerable and derivatised peptides are usually needed rather than the free amino acids and peptides, even in the simplest studies, thus somewhat defeating the object of trying to get conformational details of the peptides themselves. ^{13}C NMR spectra, involving simpler distributions of peaks, are often valuable in structural studies with longer peptides, though they are not so informative on conformational details.

Some examples illustrate the benefits of more advanced instrumentation; a ^{1}H correlated spectroscopy (COSY) NMR spectrum of a short oligopeptide (Figure 3.4) shows the 'normal NMR spectrum' across the top of the square and along the diagonal from lower left- to upper right-hand corners as it would appear through looking down on the spectrum. The 'off-diagonal' areas in the square may be interpreted to identify peaks that are difficult or impossible to assign in the 'normal' spectrum, since they may be hidden by overlapping peaks. These peaks can then be linked to particular protons in the peptide, so providing evidence of structure, including conformational details based on vicinal coupling constants giving torsion

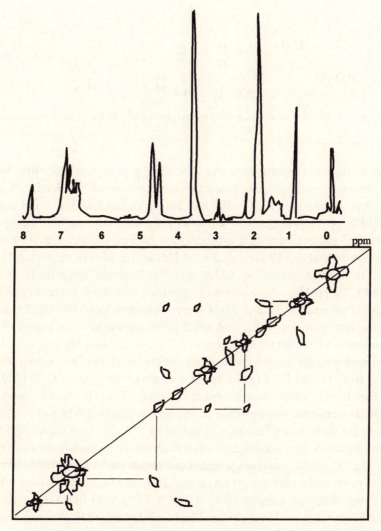

Figure 3.4. ^{1}H–^{1}H-COSY NMR spectrum of an oligopeptide containing aromatic (Phe) and aliphatic (Leu, Gly) residues.

angles; for example ϕ for the NH—C$^\alpha$H bond (Figure 2.1) within the peptide backbone, derived from the interpretation of a ^{1}H NMR spectrum.

3.8 Examples of assignments of structures to peptides from NMR spectra and other data

An L-glutamylglycine amide from the Mediterranean sponge *Achinoe tenacior* (Casapullo *et al.*, 1994) has the molecular formula $C_{15}H_{19}N_3O_5$ (M$^+$ 349 for its

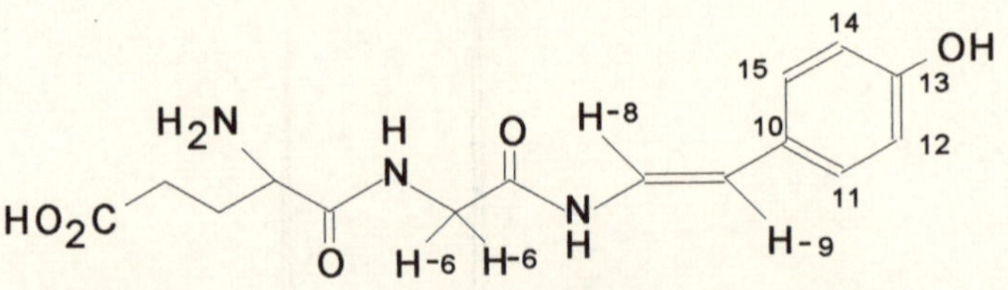

Figure 3.5. The structure of an L-glutamylglycine amide from *Achinoe tenacior*.

dimethyl derivative; the natural product failed to give a mass spectrum that could be interpreted). It was assigned a structure on the basis of NMR (^{1}H, ^{13}C, hetero-nuclear shift correlation (HETCOR) and heteronuclear multiple bond connectiv-ity (HMBC), i.e. long-range ^{1}H–^{13}C COSY), the amount available being 5.6 mg. Data revealed the presence of a *p*-substituted phenol, indicated by two mutually coupled doublets at δ 7.19 (2H, d, $J = 8.8$ Hz) and 6.74 (2H, d, $J = 8.8$ Hz), each attached to carbon atoms at 127.8 and 116.6 ppm, respectively (C^2H_3OH; HETCOR). In HMBC, both aromatic protons displayed correlations to C-13 (157.7 ppm) whereas H-12 and H-14 were correlated to C-10 (129.2 ppm). Two olefinic methine carbons (121.2 and 115.5 ppm) remained to be placed. The cou-pling constants of the attached protons (δ 7.30, d, $J = 14.6$ Hz and δ 6.21, d, $J = 14.6$ Hz) indicated the presence of a *trans* double bond. HMBC correlations of H-9 (δ 6.21) to C-11 and C-15 (127.8 ppm) and of H-8 (δ 7.30) to C-10 (129.2 ppm) linked the C=C bond to the aromatic ring. The ^{1}H NMR spectrum in $C^2H_3SOC^2H_3$ indicated the presence of one exchangeable doublet ($J = 9.8$ Hz) at δ 10.03 coupled with the olefinic proton at C-8 (δ 7.18 dd, $J = 9.8$ and 14.2 Hz) and this demonstrated the presence of the *C*-terminal *trans*-4-hydroxystyrylamine residue. The ^{13}C NMR spectrum revealed two amide carbonyl carbon atoms (168.6 and 175.5 ppm) and a carboxy group carbon atom at 185.5 ppm, whereas ^{1}H NMR revealed two amide protons (δ 10.03 d and 8.3 t in $C^2H_3SOC^2H_3$, with the two amino acids glutamic acid and glycine, identified through their characteristic chem-ical shift data (Figure 3.5).

HMBC correlation of H-8 to the carbonyl carbon atom of glycine (168.6 ppm) connected the glycine and *C*-terminal *trans*-4-hydroxystyrylamine residue. HMBC correlations of both H-6 protons to both carbonyl atoms at 168.6 and 175.5 ppm and of both H-2 protons and H-3 protons to the carboxy carbon atom at 181.5 ppm decided the structure unambiguously, the L-configuration of the glutamic acid residue being established through hydrolysis, derivatisation with Marfey's reagent and HPLC analysis (see Section 4.5.1).

Current ^{1}H and ^{13}C NMR spectroscopy techniques, with minimal help from other data, can solve problems of more complex peptide structure, as illustrated (Sano and Kaya, 1995) with a new microcystin, [D-Asp^3,Dhb^7]microcystin-RR, $C_{48}H_{73}N_{13}O_{12}$ (RMM 1024.5608 = [M + H]$^+$ obtained by high-resolution FAB mass spectrometry

Table 3.4. *^1H and ^{13}C NMR data for [D-Asp^3, Dhb^7]microcystin-RR*

Position		^1H	J (Hz)	^{13}C	Position		^1H	J (Hz)	^{13}C
Adda	1			176.7	Dhb	1			166.8
	2	3.13	(m, 7.0, 10.5)	45.0		2			132.1
	3	4.54	(dd, 8.9, 10.5)	56.7		3	5.70	(q, 7.3)	123.5
	4	5.51	(dd, 8.9, 15.6)	127.1		4	1.87	(d, 7.3)	13.4
	5	6.21	(d, 15.6)	138.6	Ala	1			175.2
	6			134.1		2	4.56	(t, 7.33)	49.7
	7	5.39	(d. 9.8)	136.7		3	1.32	(d, 7.33)	17.3
	8	2.58	(m, 9.8, 6.7)	37.7	Arg^1	1			172.0
	9	3.26	(m)	88.4		2	4.43	(m)	53.0
	10	2.81	(dd. 4.9, 14.0)	39.0		3	2.03	(m)	29.2
		2.67	(dd, 7.2, 14.0)			4	1.55	(m)	26.5
	11	1.04	(d, 7.0)	16.1		5	3.14	(m)	42.0
	12	1.61	(d, 0.9)	13.0		6			158.6
	13	0.99	(d, 6.7)	16.6	Asp	1			176.7
	14	3.23	(s)	58.7		2	4.64	(t, 4.27)	52.9
	15			140.6		3	2.90	(dd, 4.7, 13.6)	39.7
	16	7.17	(m)	130.6			2.23	(m)	
	17	7.24	(m)	129.2		4			175.1
	18	7.15	(m)	127.1	Arg^2	1			173.7
	19	7.24	(m)	129.2		2	4.21	(t, 7.5)	57.0
	20	7.17	(m)	130.6		3	2.00	(m)	29.5
Glu	1			179.5		4	1.79	(m)	26.6
	2	4.18	(dd, 8.5, 7.0)	56.3			1.72	(m)	
	3	2.08	(m)	29.5		5	3.21	(m)	42.0
		1.94	(m)			6			158.7
	4	2.47	(m)	34.3					
		2.28	(m)						
	5			175.4					

compared with the calculated value 1024.5580 for the RMM) from *Oscillatoria agardhii* (see also Section 1.13). ^1H and ^{13}C NMR data (C^{2}H$_3$O^{2}H) and structural assignments for [D-Asp^3, Dhb^7]microcystin-RR are given in Table 3.4. Distortionless enhancement by polarisation transfer (DEPT) NMR confirmed that carbon atoms at 166.8 and 132.1 ppm are quaternary carbon atoms and the rotating-frame nuclear Overhauser effect spectroscopy (ROESY) experiment (C^{2}H$_3$SOC^{2}H$_3$) showed that there was a lack of correlation between the NH and Dhb protons, whereas ^1H–^1H COSY and HMBC spectra revealed the presence of Adda (side-chain Z configurations from coupling constants and chemical shift data). The sequence (Figure 3.6) was obtained mostly by interpretation of HMBC spectra (δ-H of Adda was correlated to the carbonyl carbon of Arg-2; the methyl of Dhb was correlated to the side-

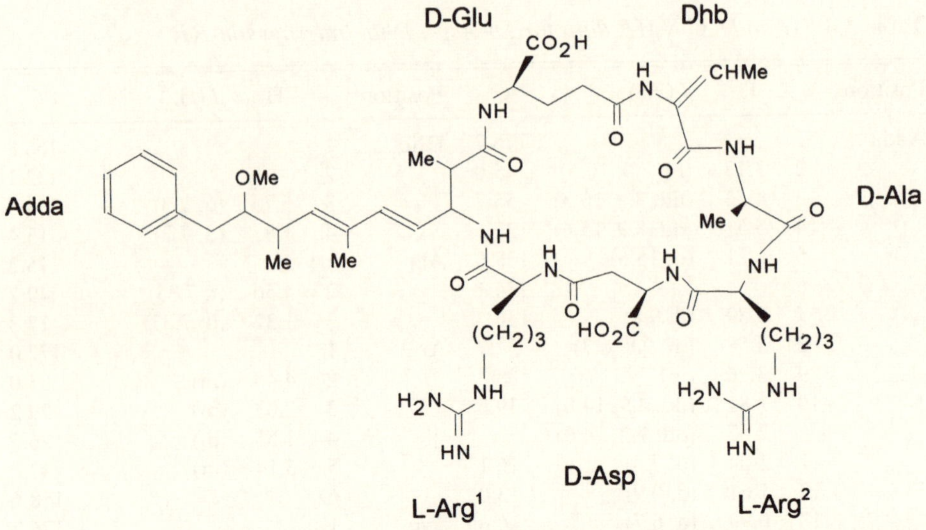

Figure 3.6. A microcystin containing 2-amino-2-butenoic acid (Dhb), isolated from *Oscillatoria agardhii*. It also contains Adda, D-Ala, D-Glu, D-Asp and two L-Arg residues. It is, strictly speaking, an isopeptide, since the aspartic and glutamic acid residues are linked through their side-chain carboxy groups.

chain carbonyl group of Glu; the connection between Dhb and Glu was confirmed by decoupled HMBC measurements).

Acid hydrolysis (6 M HCl, 16 h) and amino-acid analysis detected Ala, Asp, Glu and Arg, with Ala, Asp and Glu shown to be of D-configuration by chiral GLC analysis (on a Chirasil-L-Val column) of the mixture of *N*-trifluoroacetyl isopropyl esters obtained by derivatising the components of the hydrolysate; see Chapter 4, Part 3.

A consideration of aspects of structure elucidation for alamethicin F-30 is given in Section 4.18.2.

3.9 References

For background reading, see

Cohn, E. J. and Edsall, J. T. (1943) *Proteins, Amino Acids and Peptides as Ions and Dipolar Ions*, Reinhold Publishing Corporation, New York.
Tanford, C. and Roxby, R. (1972) *Biochemistry*, **11**, 2192 (interpretation of protein titration curves).

References cited in the text:
Casapullo, A., Minale, L. and Zollo, F. (1994) *Tetrahedron Lett.*, **35**, 2421.
Chou, P. Y. and Fasman, G. D. (1974) *Biochemistry*, **13**, 222.

Lakowicz, J. R. (Ed.) (1992) *Topics in Fluorescence Spectroscopy, Volume 3: Biochemical Applications*, Plenum Press, New York (particularly Ross, J. B. A., Laws, W. R., Rousslang, K. W. and Wyssbrod, H. R., pp. 1–63).

Miclo, L., Perrin, E., Driou, A., Mellet, M. and Linden, G. (1995) *Int. J. Pept. Protein Res.*, **46**, 186.

Palm, V. (1970) *Z. Chem.*, **10**, 31.

Sano, T. and Kaya, K. (1995) *Tetrahedron Lett.*, **36**, 8603.

Woody, R. W. (1995) *Methods Enzymol.*, **246**, 34.

4

Reactions and analytical methods for amino acids and peptides

Part 1. Reactions of amino acids and peptides

4.1 Introduction

Part 1 of this chapter is intended to provide background material for the analytical procedures described later in this chapter for amino acids and peptides, but it also provides a broad survey of the topic that can be read in isolation from the analytical context. The derivatisation of amino acids is the basis of many of the sensitive analytical amino-acid assay procedures in current use and this chapter covers the normal profile of reactions of the amino and carboxy groups, knowledge of which is an essential prerequisite for appreciating the analytical context. Reactions of peptides are also covered here (e.g. peptide and protein hydrolysis is covered in Section 4.4.7), though the coverage is restricted in scope because parts of this topic are discussed in Chapter 5, where it is relevant to sequence-determination procedures (see also Barrett, 1985).

4.2 General survey

Many reactions with amino acids also involve the side-chain functional groups and these are generally easily understood in terms of the normal profile of reactions of the functional groups concerned. Chapter 6 deals with reactions of side-chains of amino acids, since these reactions can be exploited as a way of using one amino acid to synthesise another. Also, there are often unexpected consequences owing to the involvement of side-chain functional groups (also involvement of the amide group for peptides), when a reaction is directed either at the amino or at the carboxy group of an amino acid or a peptide.

4.2.1 Pyrolysis of amino acids and peptides

Thermal breakdown of amino acids and peptides is a simple feature of their reaction behaviour that impinges on amino-acid studies. The avoidance of decomposition during the preparation of samples for analysis and some appreciation of the validity of conclusions drawn on the organic content of meteorites provide examples; on the other hand, controlled decomposition is encouraged in areas of food preparation since generation of flavour and aroma during the cooking of foods can depend to some extent on this reaction. By its nature, this is a topic for which generalisations concerning the chemistry of pyrolysis are not possible, since the side-chains of the amino acids are primarily involved and any discussion falls into a number of isolated observations.

Some breakdown, as well as cyclisation reactions, can be expected for most of the coded amino acids when they are held at temperatures around and above 200 °C. These processes lead to decarboxylation, side-chain loss to form glycine and formation of amines, furans, pyrroles and pyridines, typically. Higher temperatures (850–1000 °C) cause all the common amino acids to decompose to HCN as the major pyrolysis product, together with CO_2 and the hydrocarbon derived from the side-chain.

4.2.2 Reactions of the amino group

The most representative general reactions of the amino group (see Figure 4.1) that are, more or less, easily reversible are *acylation* (^+H_3N—$(CR^1R^2$—$)_nCO_2^- \rightarrow R$—CO—NH—$(CR^1R^2$—$)_nCO_2H$) and the analogous *sulphonylation, thioacylation* and *thiocarbamylation*. N-*Benzylation* can be reversed by catalytic hydrogenation.

Irreversible processes (e.g. *Schiff-base formation followed by reduction, leading to N-alkylation*) and modification (e.g. *guanidination*) or complete *replacement of the amino group* can be achieved e.g. by diazotisation and displacement with a number of species, such as chlorinolysis (L-α-amino acid $\rightarrow Cl$—CHR—CO_2H via the diazonium salt Cl^- $^+N_2$—CHR—CO_2H with retention of configuration; Koppenhoefer and Schurig, 1988) and analogous hydrolysis to the α-hydroxy acid. Other substitutions of NH protons, such as *N*-chlorination in water solutions by hypochlorous acid, HOCl, can be accomplished without affecting other functional groups, with most of the common α-amino acids.

4.2.3 Reactions of the carboxy group

The most representative general reactions (most of which are reversible) are: *esterification; oxidative decarboxylation; successive reduction to the aldehyde and then to the primary alcohol*; and *acyl halide formation*, giving derivatives useful for *conversion into* L-α-*acylamido-ketones* (RNH—$(CR^1R^2$—$)_nCOCl \rightarrow RNH$—$(CR^1R^2$—$)_nCOCH_3$)

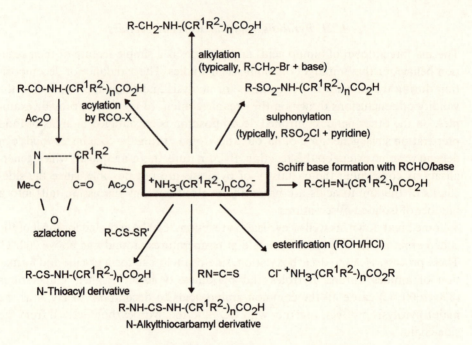

Figure 4.1. Some standard reactions of amino acids.

and *Arndt–Eistert homologation to β-amino acids* ($RNH—(CR^1R^2—)_n COCl \rightarrow RNH—(CR^1R^2—)_n CH_2CO_2H$). *Curtius rearrangement* (ester $R—CO_2R \rightarrow$ acid hydrazide $R—CONHNH_2 \rightarrow$ amine $R—NH_2$) is another example of a range of classical organic functional group transformations that can be brought about.

Quaternary ammonium salts of amino acids can be formed in the usual way (Equation (4.1); Lansbury *et al.*, 1989) and have the particular advantage that they are soluble in aprotic organic solvents (particularly the tetra-n-butylammonium salts), so opening up to amino acids (which are not significantly soluble in these solvents) a wider range of reactions (Nagase *et al.*, 1993).

$$
\begin{array}{c}
CO_2^- \\
| \\
H_3\overset{+}{N}-C-R \\
| \\
H
\end{array}
+ (alk)_4 N^+ OH^- \longrightarrow
\begin{array}{c}
CO_2^- \; N^+(alk)_4 \\
| \\
H_2N-C-R \\
| \\
H
\end{array}
+ H_2O \qquad (4.1)
$$

These routine reactions are the basis of the growing numbers of applications of natural amino acids in stereoselective synthesis (Coppola and Schuster, 1987). They are also used for the selective introduction of often exotic structures that are used as protecting groups for amino acids, giving intermediates for peptide synthesis, as illustrated in Chapter 7.

4.2.4 Reactions involving both amino and carboxy groups

These reactions are, by definition, particular to amino acids and are valuable in giving access to a wide range of heterocyclic compounds (e.g. *azlactone formation*; Figure 4.1).

4.3 A more detailed survey of reactions of the amino group

4.3.1 N-Acylation

In spite of the routine nature of the chemistry involved in N-*acylation* (Equation (4.2)), Schotten–Baumann acylation requires an acid chloride as reagent, a compound that can react with the carboxy group as well. The mystification that the consequences of this caused at the time (1962) – since dipeptides and oligopeptides were being formed in this way – actually had a constructive outcome, since the '*mixed anhydride*' procedure of peptide-bond formation (Chapter 7) was developed by the workers who unravelled the course of events (Equation (4.2)).

$$
\begin{array}{c}
\text{CO}_2^- \\
\text{+} | \quad\quad \text{NaOH} \quad\quad\quad\quad\quad\quad\quad \text{R}^1\text{COCl} \\
\text{R}^1\text{COCl} + \ \text{H}_3\text{N-C-R} \longrightarrow \text{R}^1\text{CONHCHRCO}_2\text{H} \longrightarrow (\text{R}^1\text{CONHCHRCO})_2\text{O} \\
| \quad\quad\quad\quad\quad\quad\quad \text{N-acylamino acid} \quad\quad\quad\quad \text{"symmetrical anhydride"} \\
\text{H}
\end{array}
$$

[or $\text{R}^1\text{CONHCHRCO-O-COR}^1$]
"mixed anhydride"

repetition of these steps

$$
\begin{array}{c}
\text{CO}_2\text{H} \\
| \\
\text{R}^1\text{CONHCHRCONH-C-R} \longrightarrow \quad \text{N-acyl-oligopeptides} \quad\quad\quad\quad (4.2) \\
| \\
\text{N-acyldipeptide} \quad \text{H}
\end{array}
$$

Most of the acylated and alkyloxycarbonylated amino acids, R—CO—NH—$(CR^1R^2-)_n\text{CO}_2\text{H}$ and RO—CO—NH—$(CR^1R^2-)_n\text{CO}_2\text{H}$, respectively, that are required for analytical work and for peptide synthesis, are prepared through this reaction. The preparation is carried out using methods involving additives or other specific conditions established by trial and error in order to avoid the unwanted pathways and employing leaving groups other than chloride, in many optimised procedures. Bis-acyloxycarbonylated amino acids, e.g. bis(Boc)amino acids (Boc = Bu^{t}OCO; Benoiton *et al.*, 1994) can be prepared by removal of the amide proton of a Boc-amino acid with NaH and its substitution by a second Boc group by further treatment with the acyloxycarbonylating reagent. This demonstrates that a significant level of acidity is possessed by urethanes, RO—CO—NHR, in that a proton can be removed from the grouping in this way.

Acetic anhydride reacts with an α-amino acid in the expected way, under mild conditions, to give the N-*acetyl derivative*, but also to set up an equilibrium with the carboxy group to form a *mixed anhydride*. More vigorous conditions promote the cyclisation of the mixed anhydride to the *oxazol-5(4H)-one* (the 'azlactone' in Figure 4.1), which undergoes racemisation via the oxazole tautomer under the reaction conditions. Hydrolysis at the end of the process gives the *racemised amino acid*, so the net result is useful in the conversion of a natural L-amino acid into its D-enantiomer through racemisation, followed by resolution of the racemate (Chapter 6).

4.3.2 Reactions with aldehydes

A *Schiff base* formed with an aldehyde (Figure 4.1) can be racemised readily via the azomethine ylide tautomer (cf. Equations (4.3)–(4.5)). The formol titration procedure, releasing one proton per NH_2 group through reaction of a polypeptide or protein with formaldehyde, is an obselete procedure for quantitative determination of $-NH_2$ groups, though it is sometimes still used to estimate $-CONH_2$ groups in proteins and to crosslink proteins through amide groups (Equation 4.3).

$$-CONH_2 + HCHO + NH_2CO- \longrightarrow -CO-NH-CH_2-NH-CO- \qquad (4.3)$$

$$(4.4)$$

Ruhemann's Purple

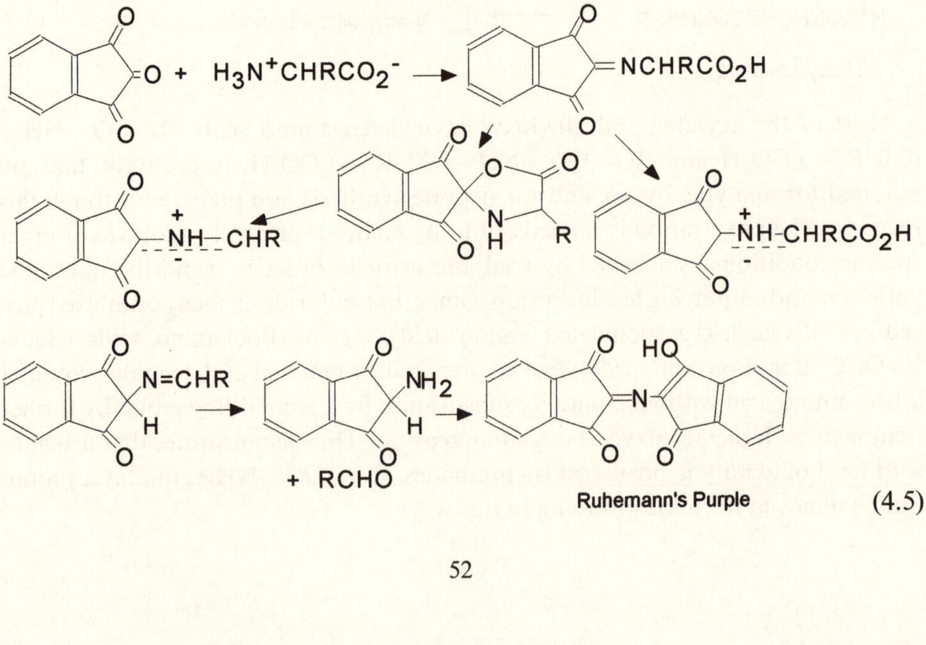

$$(4.5)$$

52

It is worth emphasising the ease of the reaction of amines and amides with aldehydes, since it explains the need to use purified solvents when dealing with amino acids, peptides and proteins, to avoid such side-reactions. Glutaraldehyde has been used for crosslinking proteins (from the earliest days in the leather industry, too) but it is toxic and therefore less in favour in laboratory work now.

The *Maillard reaction* is a long-known and complex cascade of individual reaction steps involving an amino acid and an aldose. It starts with Schiff-base formation and ensuing proton shifts within the adjacent carbohydrate chain. Nitrogen and oxygen heterocyclic products are eventually generated by reactions involving the amino group. The amino acid plus sugar starting combination ensures that this is a typical reaction that is met in food processing (it is responsible for many of the perceived enhancements to the palatability of food that cooking introduces). The reaction is also one of the factors responsible for cell 'ageing' in higher organisms (crosslinking of proteins in the living cell; Grandhee and Monnier, 1991).

As well as the amino and carboxy groups, the amino-acid side-chain is also involved early in the Maillard reaction, so the side-chain functional groups become incorporated into the products from Maillard reactions. Cysteine, in particular, reacts with glucose to generate numerous reaction products, some of which introduce attractive tastes and aromas to foods associated with sulphur functional groups.

Appreciation of the importance of this process in enhancing the palatability of food has led to the industrial synthesis of Maillard products that feature as synthetic food additives appealing to modern tastes. Some of these have been used in highly flavoured snack foods.

4.3.3 N-*Alkylation*

Control of the degree of alkylation of an amino acid is difficult. Alkylation can lead to *N*-mono- and di-alkyl derivatives, or betaines $R_3^1N^+$—$(CR^1R^2$—$)_nCO_2^-$, when an alkyl halide is used. Many practical devices can be employed to get the mono-alkylated compound. Schiff-base formation with an aldehyde, followed by reduction, is a standard route.

4.4 A survey of reactions of the carboxy group

Although the carboxy group undergoes many of the reactions expected of it, without affecting the amino group, there are some reactions that bring about modification of both functional groups. An example already mentioned is the formation of an oxazol-5(4H)-one (an 'azlactone'; Figure 4.1). This illustrates one of the many uses of amino acids in heterocyclic synthesis.

4.4.1 Esterification

Fischer–Speier esterification to give a salt of an amino acid ester (by refluxing an alkanol with anhydrous HCl or hot benzyl alcohol with toluene-p-sulphonic acid) is straightforward. The fact that the nearby amino group is protonated when the carboxy group reacts to give α-amino acid esters does not slow the reaction down unduly; a nearby positive site might have been expected to reduce the electrophilic character of the carboxy carbon atom. Without an acid catalyst, N-alkylation can accompany esterification (Equation 4.6).

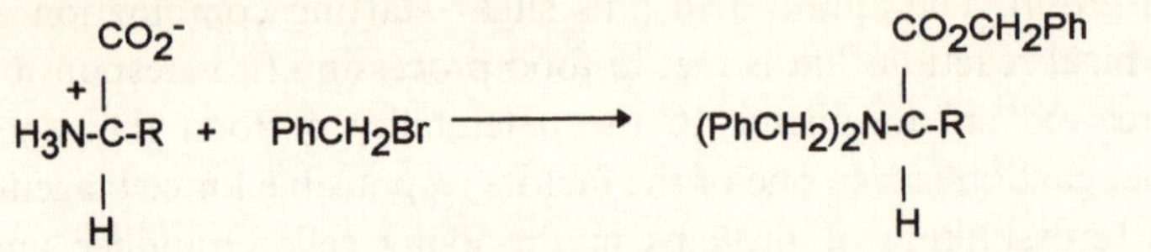

$$
\begin{array}{c}
CO_2^- \\
+| \\
H_3N\text{-}C\text{-}R \\
| \\
H
\end{array}
+ \; PhCH_2Br \longrightarrow
\begin{array}{c}
CO_2CH_2Ph \\
| \\
(PhCH_2)_2N\text{-}C\text{-}R \\
| \\
H
\end{array}
\qquad (4.6)
$$

N-acylated amino acids that are acid-sensitive (as are many of the N-protected amino acids used in peptide synthesis) are converted into esters through other means, commonly through the use of N,N'-dicyclohexylcarbodiimide and an alcohol or a phenol, especially for the preparation of 'active esters' for use in peptide synthesis (Chapter 7).

Diazomethane is a useful reagent for the conversion of carboxy groups into methyl esters. It is especially convenient since the yellow colour of the reagent disappears as it is used up, thus providing a convenient indication of the progress of the reaction.

4.4.2 Oxidative decarboxylation

Strecker degradation (Schonberg and Moubasher, 1952) is a characteristic reaction of α-amino acids, the well-known ninhydrin colour reaction being a classic example of the process. Only very recently has the nature of the reactive azomethine ylide intermediate been assigned correctly (Equation (4.5); Grigg *et al.*, 1989). The dipole is formed first from a Schiff base and the oxazolidin-5-one formed from this suffers decarboxylation to form another Schiff base, hydrolysis giving an aldehyde and 2-amino-indan-1,3-dione (which rapidly condenses with another molecule of ninhydrin to give Ruhemann's Purple). Overall, an α-amino acid $^+NH_3$—CR^1R^2—CO_2^- is degraded into CO_2, RCHO (when $R^1 = R$, $R^2 = H$) and NH_3, but α-imino acids (proline, etc.) and β-, γ- . . . amino acids do not react, so these higher homologous amino acids do not give the blue colour with ninhydrin.

4.4.3 Reduction

Successive reduction to α-amino aldehydes (2-amino-alkanals) and to 2-amino-alkanols is relatively easily managed using hydride-based reagents, once one has established the appropriate hydride reagent and protocol (Jurczak and Golebiowski, 1989). These are valuable chiral materials, since the aldehydes can be readily elaborated through aldol formation, into β-hydroxy-γ-amino acids such as statine, a constituent of the peptide antibiotic bestatin, and used in syntheses of similar amino acids in HIV protease inhibitors.

4.4.4 Halogenation

Aqueous sodium hypochlorite causes Strecker degradation of α-amino acids through initial chlorination and then oxidative imine formation, as described in Section 4.4.6. Free-radical bromination (by *N*-bromosuccinimide) of aliphatic *N*-benzoyl or *N*-phthaloyl amino acids (those without side-chain functional groups) results in side-chain (β-, γ-substitution, etc.) rather than α-substitution, the *N*-protecting group controlling the ease of reaction (Easton *et al.*, 1989). L-Isoleucine in trifluoroacetic acid solution undergoes free-radical γ-substitution with chlorine under UV irradiation and the product can be cyclised to give *trans*-3-methyl-L-proline when the reaction mixture is made alkaline (Equation (4.7)). Iodination of the aryl moiety of aromatic side-chains is mentioned in Chapter 8.

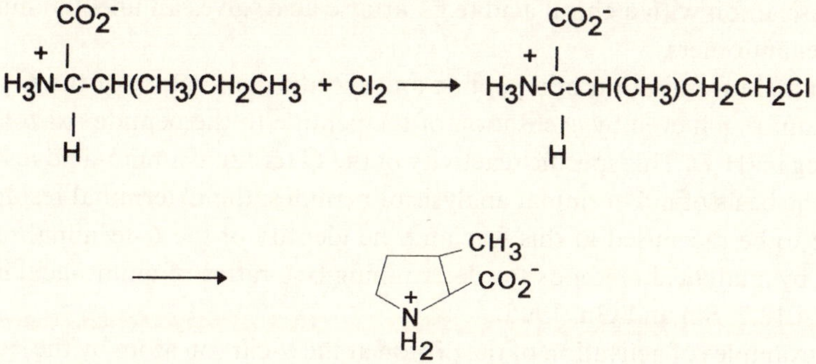

$$
\underset{H}{\overset{+}{H_3N}}\text{-C-CH(CH}_3\text{)CH}_2\text{CH}_3 + Cl_2 \longrightarrow \underset{H}{\overset{+}{H_3N}}\text{-C-CH(CH}_3\text{)CH}_2\text{CH}_2\text{Cl}
$$

(4.7)

4.4.5 Reactions involving amino and carboxy groups of α-amino acids and their N-acyl derivatives

These can include self-condensation to give peptides (in the case of the free amino acids) and cyclisation of amino-acid derivatives to give heterocyclic compounds. Controlled peptide synthesis is discussed later (Chapter 7), but the extraordinary ease of self-condensation of amino acids in concentrated aqueous NaCl containing copper(II) salts deserves mention (Saetia *et al.*, 1993) since it suggests a way through which pre-biotic peptide synthesis may have come about.

4.4.6 Reactions at the α-carbon atom and racemisation of α-amino acids

The α-carbon atom is the chiral centre of α-amino acids, which are all homochiral except glycine and its symmetrical αα-di-alkyl analogues, and reactions that generate a carbanion at this site may be accompanied by racemisation. The resistance of α-amino acids to racemisation when they are taken through many standard reactions indicates the difficulty of achieving de-protonation, though changes at the amino and carboxy groups and certain side-chain features can attenuate the reactivity at the α-carbon atom. Thus, Schiff bases can be alkylated via the di- and tri-anions (Equation (4.8)).

$$
\begin{array}{c}
CO_2^- \\
\overset{+}{H_3N}\text{-}C\text{-}R + RCHO \longrightarrow R\text{-}CH{=}N\text{-}CHR\text{-}CO_2H \\
| \quad\quad\quad -2\,H^+ \\
H
\end{array}
$$

$$
R\text{-}CH{=}N\text{-}RC^-\text{-}CO_2^- \longrightarrow R\text{-}CH{=}N\text{-}CR_2\text{-}CO_2H \tag{4.8}
$$

Acetylation and ensuing cyclisation to the oxazolone also activates the α-proton (Figure 4.1) and provides the classical route from a natural L-α-amino acid to its DL form through hydrolysis of the derived oxazolone; however, the re-protonation of a Schiff-base anion with a chiral acid (e.g. tartaric acid) gives an unequal mixture of D and L enantiomers.

C-terminal racemisation of a peptide and specific deuteration of the C-terminal residue can be achieved by cyclisation of the peptide to the peptide oxazolone and quenching in ^2H_2O. This specific reactivity of the C-terminal amino-acid residue has formed the basis of a C-terminal analysis of peptides; the C-terminal residue is the only one to be racemised in this way and the identity of the C-terminal residue is revealed by analytical methods for determining D:L ratios of amino-acid mixtures (Section 4.18.2; Sih and Gu, 1995).

Clear examples of activation of the proton at the α-carbon atom by the side-chain include cysteine, which, in the form of its N-Fmoc-S-trityl derivative, will racemise with organic base (di-isopropylethylamine) during attempts to build it into peptides, even though it is chirally stable in neutral aqueous media (Kaiser et al., 1996). It is difficult to explain this effect of a β-placed sulphur atom and it is also difficult to explain why lysine is easily racemised in aqueous solutions by irradiation with light of wavelengths shorter than 300 nm, when cadmium(II) sulphide particles are present (Ohtani et al., 1995). However, the racemisation in base of phenylglycine (H_3N^+—CHPh—CO_2^-) and bond cleavages to give benzaldehyde when it is dissolved in aqueous NaOH are more easily understood since the α-carbon atom in this amino acid will have the well-known enhanced reaction profile of a benzylic carbon atom.

A useful synthetic manipulation at the α-carbon atom of some imino acids, which does not rely on the presence of neighbouring activating groups, is anodic α-methoxylation in methanol (Shono et al., 1984). The resulting aminal has a reactive methoxy group that can be substituted by nucleophiles under mild conditions. L-Proline is a useful chiral synthon and its anodic α-methoxylation reaction (in which stereochemical integrity is maintained) has been used for alkaloid synthesis.

Mild oxidation of the *C*-terminus of a peptide can bring about the formation of an acylimide, which is then easily hydrolysed. This change, which is thought to involve oxidative α-hydroxylation of the *C*-terminal amino acid residue (or oxidation of the oxazolone formed from the *C*-terminal amino acid residue) has been accomplished under physiological conditions in the absence of enzymes when the *C*-terminus is 'activated' as an anhydride or as an oxazolone (the reactions are shown in Scheme 8.3; Barrett *et al.*, 1978) and the search for an enzymic equivalent has led to the discovery of a family of amidating enzymes. This, then, is how a biologically inactive 'propeptide', with one amino-acid residue more than the peptide amide into which it is processed, is a latent precursor for many hormones that are peptide amides (calcitonin, vasopressin, etc.).

4.4.7 Reactions of the amide group in acylamino acids and peptides

Hydrolysis of the amide bond is the best-known reaction of this functional group, in the biological context (digestion of proteins by proteinases) as well as in the organic chemical context (aqueous hydrolysis in 6 M hydrochloric acid for 12 h at 120 °C or by dilute alkali). However, the essential role of a catalyst is made clear by the fact that a peptide dissolved in *pure water* survives unchanged for many months, even under reflux.

There are numerous protocols for protein hydrolysis, involving minor variants of the standard procedure, that are intended to minimise the destruction of particular amino acids (tryptophan and cysteine/cystine in particular) through the sensitivity of their side-chains to the reaction conditions, especially when access of oxygen is not prevented. Tryptophan largely survives alkaline hydrolysis (but other coded amino acids, particularly serine and threonine, but also arginine and cysteine, do not).

Partial hydrolysis by aqueous acid is a regioselective amide-cleavage process, although it does not relate to a particular amide bond; the most easily hydrolysed amide bond in a polyamide is the one that is most exposed to reagents or otherwise enhanced in its propensity to hydrolysis. Partial hydrolysis was an important feature of the earliest structure determinations for peptides (e.g., by Sanger; see Chapter 5), and currently features in a method for mass-spectrometric structure determination of peptides (Section 4.11). The alkylation of the amide bond in peptides, described in Chapter 5, assists mass-spectrometric study through increasing the volatility of peptides.

The Edman sequencing reaction of peptides and proteins exploits the regioselective cleavage of an amide group (that amide group involving the *N*-terminal amino-acid residue) and is brought about through *N*-terminal phenylthiocarbamylation

(Figure 4.1) followed by acid-promoted cleavage (Chapter 5). Specific chemical cleavage of the backbone of peptides, which is an alternative way of breaking long polypeptide chains into smaller peptides, depending upon participation by certain side-chain functional groups, is discussed in Chapter 5.

4.5 Derivatisation of amino acids for analysis

There are two main objectives under this heading: (i) *to convert amino acids into volatile derivatives for GLC and mass-spectrometric study* and (ii) *to introduce a group or groups with conveniently measured ultraviolet/visible absorption or fluorescence characteristics*, so that *TLC, HPLC, circular dichroism and other physicochemical and spectroscopic studies, or analytical separation of amino acid mixtures*, may be accomplished. The analytical methods themselves are discussed in Chapter 3 and in later sections of this chapter.

4.5.1 *Preparation of* N-*acylamino acid esters and similar derivatives for analysis*

GLC and GC–MS analysis of amino acids requires efficient procedures for the *N*-acylation and esterification of an amino acid. Most users have settled for the preparation of *N*-trifluoroacetyl or *N*-heptafluorobutyroyl derivatives of amino-acid n-butyl esters. The earlier trend in favour of simultaneous trimethylsilylation of amino and carboxy functional groups (as well as side-chain oxygen and nitrogen functions) to prepare suitable derivatives has waned in popularity, although it may come back into favour because the t-butyldimethylsilyl groups is more stable against water and is being used increasingly frequently.

Fluorescent and UV-absorbing tagging groups include 1-(*N*,*N*-dimethylamino)-naphthalene-5-sulphonyl ('dansyl'), 9-fluorenylmethoxycarbonyl ('Fmoc'; Carlton and Morgan, 1989) and fluoresceamine and iso-indolyl (*o*-phthaldialdehyde (OPA)–β-mercaptoethanol) condensation products (Figure 4.2). All these have been used for 'pre-column derivatisation' of amino acids for high-performance liquid chromatography (HPLC), allowing relatively inexpensive amino-acid analysis based on general-purpose laboratory equipment.

An even simpler protocol is employed in the formation of *N*-phenylthiocar-bamoyl derivatives (PTC-amino acids, $C_6H_5NHCSNHCHRCO_2H$; Figure 4.1) by reaction in a suitable buffer with the Edman reagent, phenyl isothiocyanate (West and Crabb, 1989).

Unfortunately, the OPA condensation products decompose to some extent on the HPLC column (as shown by means of ^{14}C-labelled amino acids – the positions of emerging samples shown by their fluorescence lag behind maximum radioactivity profiles), although users favouring these OPA derivatives have learned to practise strict protocols for their use, in order to get reliable results. The OPA–*N*-acetyl-L-cysteine condensation product is a diastereoisomer mixture when formed with a

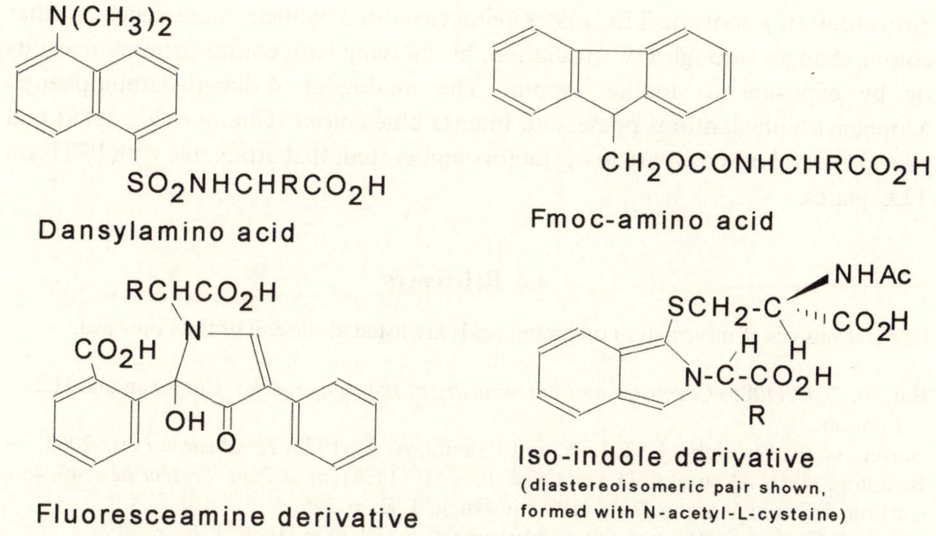

N(CH$_3$)$_2$

SO$_2$NHCHRCO$_2$H

Dansylamino acid

CH$_2$OCONHCHRCO$_2$H

Fmoc-amino acid

RCHCO$_2$H

CO$_2$H

N

OH O

Fluoresceamine derivative

SCH$_2$-C

NHAc

CO$_2$H

H H

N-C-CO$_2$H

R

Iso-indole derivative

(diastereoisomeric pair shown,
formed with N-acetyl-L-cysteine)

Figure 4.2. Structures of some derivatised amino acids.

racemic or partly racemised amino acid and quantitative HPLC provides the enantiomer ratio directly (Duchateau *et al.*, 1989).

Diastereoisomer-forming derivatisation (an example is shown in Figure 4.2), as an alternative to the separation of any of the derivatives shown in Figure 4.2, based on the use of a chiral stationary phase for HPLC or TLC, leads to quantitative enantiomer ratio determination. The earlier approach, chromatographic separation after derivatisation of an amino acid of unknown absolute configuration, or unknown enantiomeric composition, using Marfey's reagent, *N*-[5-(1-fluoro-2,4-dinitrophenyl)]-L-alaninamide (Marfey, 1984), is reminiscent of the chemistry that underpinned Sanger's sequence determination of proteins based on *N*-(2,4-dinitrophenyl)ation. With HPLC methodology, the method continues to be used because it is consistent in giving good separation of diastereoisomers and the L,L diastereoisomer consistently elutes before the D,L isomer does. A similar approach, but one that is particularly suited to NMR spectroscopic determination (^{1}H, ^{13}C and ^{19}F variants) of diastereoisomer ratios, uses Mosher's reagent, $(-)$-(S)-α-methoxy-α-trifluoromethyl-α-phenylacetic acid anhydride to acylate the amino group of an amino acid and the resulting mixture is assayed (Dale *et al.*, 1969).

Thin-layer chromatography (TLC) provides convenient routine analytical support of synthesis and other amino-acid interests and has been used in the Mosher procedure just described. It is most generally used for free amino acids and peptides, with spray reagents based on ninhydrin, or on the above derivatives ('post-TLC derivatisation'). Dansyl and phenylthiohydantoin (PTH) derivatives have been used for many years for identifying amino acids in mixtures by TLC ('pre-TLC derivatisation'), spots on TLC plates being visualised by their fluorescence or other colour changes through UV irradiation, by spraying with colour-forming reagents or by exposure to iodine vapour. The analogous 4-dimethylaminophenyl-azophenylthiohydantoins possess an intense blue colour (Chang *et al.*, 1989) and provide a level of sensitivity many factors higher than that attainable with PTHs on TLC plates.

4.6 References

General sources of information on amino acids are listed at the end of the Foreword.

Barrett, G. C. (1985) *Chemistry and Biochemistry of the Amino Acids*, Chapman and Hall, London.

Barrett, G. C., Chowdhury, M. L. A. and Usmani, A. A. (1978) *Tetrahedron Lett.*, 2063.

Benoiton, N. L., Akyuvekli, D. and Chen, F. M. F. (1994) *Int. J. Pept. Protein Res.*, **45**, 466.

Carlton, J. E. and Morgan, W. T. (1989) in Hugli, T. E., p. 266.

Chang, J. Y., Knecht, R., Jenoe, P. and Vekemans, S. (1989) in Hugli, T. E., p. 305.

Coppola, G. M. and Schuster, H. F. (1987) *Asymmetric Synthesis: Construction of Chiral Molecules using Amino Acids*, Wiley, New York.

Dale, J. A., Dull, D. L. and Mosher, H. S. (1969) *J. Org. Chem.*, **34**, 2543.

Duchateau, A., Crombach, M., Kemphuis, J., Boestgen, W. H. J., Schoemaker, H. E. and Meijer, E. M. (1989) *J. Chromatogr.*, **471**, 263.

Easton, C. J., Tan, E. W. and Hay, M. P. (1989) *J. Chem. Soc., Chem. Commun.*, 385.

Grandhee, S. K. and Monnier, V. (1991) *J. Biol. Chem.*, **266**, 11 649.

Grigg, R., Malone, J. F., Mongkolaussararatana, T. and Thianpatanagul, S. (1989) *Tetrahedron*, **45**, 3849.

Hugli, T. E. (1989) *Techniques of Protein Chemistry*, Academic Press, San Diego, California.

Jurczak, J. and Golebiowski, A. (1989) *Chem. Rev.*, **89**, 1197.

Kaiser, T., Nicholson, G. J., Kohlbau, H. J. and Voelter, W. (1996) *Tetrahedron Lett.*, **37**, 1187.

Koppenhoefer, B. and Schurig, V. (1988) *Org. Synth.*, **66**, 151.

Lansbury, P. T., Hendrix, J. C. and Coffman, A. I. (1989) *Tetrahedron Lett.*, **30**, 4915.

Marfey, P. (1984) *Carlsberg Res. Commun.*, **49**, 591.

Nagase, T., Fukami, T., Urakawa, Y., Kumagai, U. and Ishikawa, K. (1993) *Tetrahedron Lett.*, **34**, 2495.

Ohtani, B., Karaguchi, J., Kozowa, M., Nichimoto, S., Inui, T. and Izawa, K. (1995) *J. Chem. Soc., Faraday Trans.*, **91**, 1103.

Saetia, S., Liedl, K. R., Eder, A. H. and Rode, B. M. (1993) *Origins Life Evol. Biosphere*, **23**, 177.

Schonberg, A. and Moubasher, R. (1952) *Chem. Rev.*, **50**, 261.

Shono, T., Matsumura, Y. and Tsubata, T. (1984) *Org. Synth.*, **63**, 206.

Sih, C. J. and Gu, Q.-M. (1995) *Int. J. Pept. Protein Res.*, **46**, 366.

West, T. E. and Crabb, J. W., in Hugli, T. E. (1989) p. 295.

Part 2. Mass spectrometry in amino-acid and peptide analysis and in peptide sequence determination

4.7 General considerations

Amino acids and peptides have been considered to be difficult to study by mass spectrometry (MS) except by the more sophisticated modern instrumental techniques, though *derivatised amino acids and peptides* are readily analysed using routine laboratory spectrometers. The spectra can also give useful information, particularly through GLC–MS analysis (and recently through *capillary zone electrophoresis CZE–MS)*; see Section 4.17.1) of mixtures of amino acids and peptides (see also Section 4.11.1).

The mass spectrum of an organic compound can differ in minor details, when comparing spectra from one mass spectrometer with those obtained with another (even for instruments using the same means of ionisation). Examples of spectra given in this chapter should be viewed in this light if comparisons with published compilations of spectra (e.g., Desiderio, 1991) are made. Accurate mass values obtained by high-resolution MS can provide crucial help in structure determination of complex peptides.

4.7.1 Mass spectra of free amino acids

The difficulty in studying free amino acids by MS is their low volatility. The simplest mass spectrometers call for samples with a sufficient vapour pressure, which can often be attained with intractable samples by raising the temperature of the sample-inlet system. The trade-off when high-temperature ionisation is used is that against the thermal stability of the sample. This is the source of the problem, since extensive thermal degradation of free amino acids occurs when the simplest mass spectrometers are used with ion sources at high temperatures.

However, advances in instrumentation have provided mass spectra of all twenty common amino acids in their un-derivatised form (Bouchonnet *et al.*, 1992). Plasma desorption MS with electrospray ionisation (Section 4.11.1) has been used to overcome the problems of low volatility. The amino acids give positive ions through protonation in this technique ($M \rightarrow MH^+$), which then fragment with loss of 46 atomic mass units and so give immonium ions $[H_2N{=}CHR]^+$, all of which (except the ion from arginine) undergo four successive fragmentations in the mass spectrometer before giving ions that are registered as the mass spectra.

The sort of information that more sophisticated instrumentation has provided relates to subtle structural details of free amino acids, such as intramolecular hydrogen bonding and intermolecular associations in the gas phase. These details are not really relevant to analytical work and are not covered here.

61

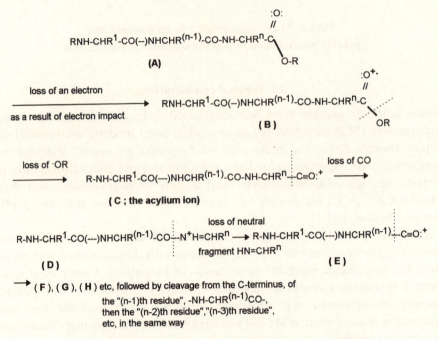

Scheme 4.1. Fragmentation of a peptide after electron impact.

4.7.2 Mass spectra of free peptides

For the same reasons, free peptides are also less amenable to mass-spectrometric study than are peptide derivatives, but some peptides are sufficiently volatile and can be ionised in the gas phase by electron impact in the mass spectrometer. The molecular ions formed in this way are rapidly fragmented in a characteristic fashion, as a result of the stepwise expulsion of amino-acid residue units from the *C*-terminal end of the chain, as shown by the sequence of structures **A** to **E** in Scheme 4.1.

The interpretation of routine electron-impact mass spectra (EIMS) illustrated here for peptides (Johnstone and Rose, 1983) starts with the identification of a positively charged site in a molecule that has suffered electron impact. The atom of lowest ionisation potential in the molecule (the atom from which an electron is most easily lost) becomes the positive site, creating the 'parent ion', M^+, otherwise called 'the molecular ion' (**B** in Figure 4.3). The oxygen atom of a carbonyl group in a carboxylic acid or ester has the lowest ionisation potential among common functional groups and the *C*-terminal carbonyl group of a peptide is therefore the site at which a positive charge develops preferentially in a peptide derivative.

When a positive site has developed in a molecule through electron impact, bond cleavages (e.g. $A \rightarrow B \rightarrow C \rightarrow D \rightarrow E$ in Scheme 4.1) are initiated within a very short time, creating 'daughter ions' (m_1^+, m_2^+, etc.) Only positive ions are recorded as the mass spectrum of the sample in standard mass-spectrometric analysis and, since

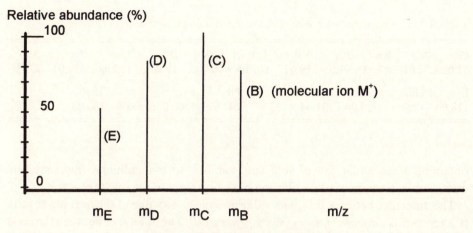

Figure 4.3. A partial mass spectrum.

these ions are created by fragmentation from the *C*-terminus of a peptide, the mass spectrum can be interpreted to reveal the sequence of a peptide. Examples of sequence determination by interpretation of mass spectra are given in the following sections. In this process

(a) the carboxy-terminus, $-CO-NH-CHR^n-CO_2R$, of the derivatised peptide $R-NH-CHR^1-CO(...)NHCHR^{(n-1)}-CO-NH-CHR^n-CO_2R$ **(A)** suffers ionisation as a result of electron impact and is stripped of one electron from the *C*-terminal carboxy group, giving the molecular ion M^+ **(B)**, which then loses the radical $\cdot OR$, to become $R-NH-CHR^1-CO(...)NHCHR^{(n-1)}-CO-NH-CHR^n-C\equiv O:^{\cdot+}$ **(C)**; and

(b) **(C)** loses CO by fragmenting to $R-NH-CHR^1-CO(...)NHCHR^{(n-1)}-CO-N^+H=CHR^n$ **(D)**, which then loses $HN=CHR^n$ to give $R-NH-CHR^1-CO(...)NHCHR^{(n-1)}-C\equiv O:^{\cdot+}$ **(E)**.

The process is repeated at the $-C=O$ group of the newly exposed *C*-terminus of the peptide and continues in the same way towards the *N*-terminus, sequential loss of residues from the carboxy-terminus creating one new positive ion at each new fragmentation. The mass spectrum produced for this peptide, *if this were the only way in which the peptide is fragmented and if all the positive ions B–E survive long enough to reach the ion collector in the mass spectrometer*, is shown in Figure 4.3.

The mass value of the peak of highest mass (**(B)** in Scheme 4.1) can be assumed, as a start to interpreting the mass spectrum of an 'unknown' peptide, to be the molecular mass; in other words, *(B) is the molecular ion M^+ and its mass value is the molecular weight of the peptide*. This assumption is not always valid for EIMS, but other variants of MS (see later) are particularly useful in maximising the chance of

Table 4.1. *The relative molecular mass of each 'coded' amino-acid residue*

Gly	Ala	Ser	Pro	Val	Thr	Cys	Ile	Leu	Asn	Asp
57.02	71.04	87.03	97.05	99.07	101.05	103.01	113.08	113.08	114.04	115.03

Lys	Gln	Glu	Met	His	Phe	Arg	Tyr	Trp
128.09	128.06	129.04	131.04	137.06	147.07	156.10	163.06	186.08

obtaining a molecular ion of sufficient stability, so that minimal fragmentation occurs and a prominent molecular ion peak is present in the mass spectrum.

The difference between this mass value and the mass value of the next lower peak (**C**), represents *the mass of the esterifying group OR*. The difference between the mass value of (**C**) and that of the next lower mass peak (**D**), represents *the mass of CO* (28 atomic mass units). The difference between the mass value of (**C**) and that of (**E**) represents *the mass of the C-terminal amino acid residue of the original peptide*, whereas the difference between (**D**) and (**E**) is the mass of the imine HN=CHR.

These mass differences for —NH—CHR^n—CO—, the residue of each of the coded amino acids built into peptides, are shown in Table 4.1 (in order of increasing molecular mass). There are actually several ways (six ways, in all; a, b, c, x, y and z in Figure 4.4) in which fragmentation of the peptide backbone can occur within the molecular ion of a peptide. As well as these cleavages, the peptide may suffer (a) proton shifts accompanying skeletal rearrangements of aliphatic side-chains that can occur within the few microseconds needed for ionisation, then fragmentation, then arrival at the ion collector in the mass spectrometer; and (b) loss of complete side-chains without any cleavage of the backbone of the peptide (peaks in the high-mass end of the spectrum = peak at ([M + H]$^+$ − mass of side-chain).

As with all mass spectra, small peaks will be present due to the stable isotopes (^{13}C in particular, one or more atomic mass units to higher mass of the major peaks, with intensities up to about 10% of the intensity of the adjacent major peak). Therefore, the mass spectrum of a peptide typically has many more peaks than are present in the idealised mass spectrum shown in Figure 4.3. Some of these can indicate the presence of certain amino acids, but they indicate nothing about the sequence; thus, at the low-mass end, peaks due to individual amino acids that have been protonated and have lost CO, giving [H$_2$N=CHR]$^+$ through cleavage mode 'a', may be identified (e.g. mass 72 may indicate valine; mass 120 may indicate phenyl-alanine, etc.).

Most of the other peaks are indicative of the sequence, on the basis of the mass differences between them. Some of these peaks may be recognised to form a series and there may be several such series corresponding to the various cleavage modes (a, b, c, x, y and z; Figure 4.4). Members of a particular series will show mass differ-

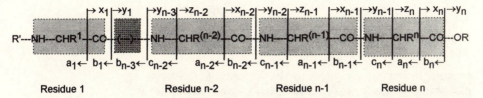

Figure 4.4. 'a_n' and 'x_n' correspond to the backbone cleavage mode shown in Scheme 4.1; thus cleavage mode 'a' gives ion D and neutral fragment CO, whereas cleavage 'x' would give the positive ion CO^+ and neutral fragment D. Arrows point in the direction of the structure that carries the positive charge and this is the part of the structure whose mass is seen as a peak in the mass spectrum.

ences between consecutive peaks (starting from the high-mass end and moving towards lower mass) that reveal the sequence.

4.7.3 Negative-ion mass spectrometry

Gas-phase ionisation by electron impact (and by other means, see later) generates many more positive ions than negative ions and conventional EIMS measurements therefore concentrate on the positive ions. Newer mass spectrometers offer the option of negative-ion EIMS, which can have some advantages such as 'cleaner' spectra (less 'background' – fewer peaks near the baseline) and intense $[M-1]^-$ peaks.

4.8 Examples of mass spectra of peptides

4.8.1 Electron-impact mass spectra (EIMS) of peptide derivatives

Mass-spectral interpretation can be illustrated (Figure 4.5) for a simple peptide derivative, N-trifluoroacetyl-valyl-glycyl-alanine methyl ester. The structure of N-TFA—Val—Gly—Ala—OMe in Figure 4.5 is labelled to show the cleavage points by vertical dashed lines, the positive ion formed by each cleavage being indicated by the direction of the arrow at the bottom of each dashed line. Bond cleavage places the positive charge on the N-terminal fragment (by convention the N-terminus is shown on the left-hand side of the structural formula of a peptide).

In developing the example of the mass spectrum of N-TFA—Val—Gly—Ala—OMe and in the interpretation of other mass spectra later in this chapter, it will be seen that the interpretation of a mass spectrum of a peptide relies on this consistent manner of bond cleavage. The electron-impact mass spectrum (EIMS) of N-TFA—Val—Gly—Ala—OMe shown in Figure 4.5 contains a molecular ion M^+ at m/z 355 and a peak at m/z 324, proving the occurrence of the primary cleavage process $A \to B \to C$ illustrated in Scheme 4.1, since a loss of 31 atomic mass units from the intact

65

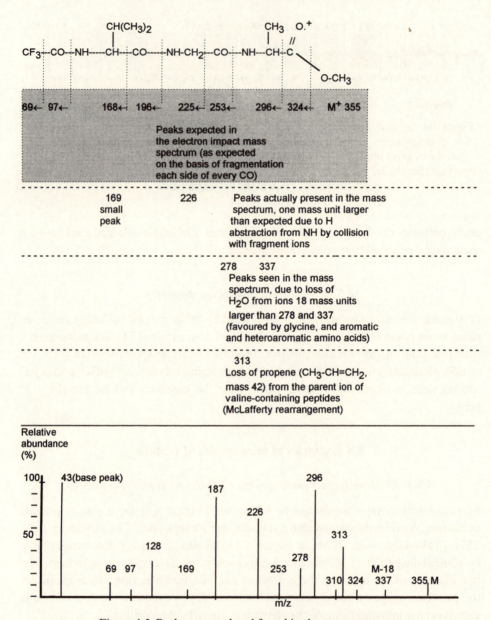

Figure 4.5. Peaks expected and found in the mass spectrum.

molecule can be explained best on this basis. Loss of 28 (CO) and then 43 atomic mass units is represented by the presence of ions at m/z 296 and m/z 253. The overall loss from the intact molecule of 102 atomic mass units establishes the C-terminal amino acid residue to be alanine. The peak at m/z 296 is the most prominent within the mass range beyond $m/z \simeq 50$ and is second in intensity only to the base peak at

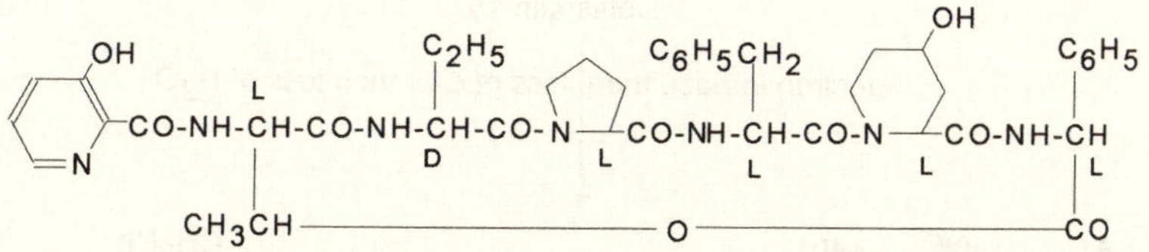

Figure 4.6. Staphylomycin S$_2$ (alias Virginiamycin S$_1$).

m/z 43. This example demonstrates the over-riding preference for ionised peptide derivatives to undergo cleavage *in the* C-*terminal region* to form two cleavage products with the *N*-terminal cleavage product being the positively charged ion and therefore the only one appearing in the mass spectrum.

Further peaks in the mass spectrum of *N*-TFA—Val—Gly—Ala—OMe can be assigned by following the same reasoning, since they are the result of cleavages on one side or the other of every carbonyl group, with retention of the positive charge by the *N*-terminal fragment after C—C or C—N bond cleavage. The sequence is thus shown to be glycine next to *C*-terminal alanine and valine next to glycine.

Staphylomycin S$_2$ The example just worked through amounts to a confirmation of a known structure, but a real application to structure assignment is illustrated by Staphylomycin S$_2$ (Figure 4.6; Vanderhaege and Parmentier, 1971). Although a structure was assigned to this cyclic peptide antibiotic through chemical degradation methods, the ease of structure assignment through mass spectra alone is notable (Compernolle *et al.*, 1972). The carbonyl group of the lactone (cyclic ester) is the site of ionisation, just as it is with the ester grouping of *N*-TFA—Val—Gly—Ala—OMe discussed above (Figure 4.5), and fragmentation follows the usual course from the molecular ion, which first loses CO and then H$_2$O (see also the next example).

Dolastatin 15   Mass-spectrometric analysis is well-suited to the study of samples that are accessible in microscopic quantities only. An example of this type, which was solved by MS, is Dolastatin 15, from the Indian Ocean sea hare *Dolabella auricularia*). It is a strongly cytostatic depsipeptide and therefore of considerable interest in leukaemia therapy (Pettit *et al.*, 1989). Routine identification of its hydrolysis products showed that it must be the ester analogue of a heptapeptide with one ester bond in place of one amide bond.

The presence of an *N*-alkylated terminal amino-acid residue (*N,N*-dimethyl-valine), deduced from the mass spectrum, means that chemical sequencing (Edman degradation) is ruled out. The sequence followed from the interpretation of the mass spectrum. The usual site of ionisation, the *C*-terminal carbonyl group, was the starting point for initial speculation on interpretation of the mass spectrum to solve the sequence. The positions of valine, *N*-methylvaline, proline, 'hydroxyvaline' (i.e. 2-

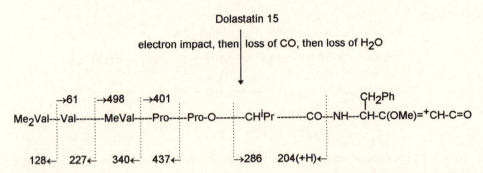

Figure 4.7. Fragmentation of Dolastatin 15 after ionisation in the gas phase in the mass spectrometer.

hydroxy-3-methylbutanoic acid) and the hitherto unknown modified phenylalanine *C*-terminal residue followed unambiguously (Figure 4.7).

4.8.2 *Finer details of mass spectra of peptides*

Further assistance in the interpretation of mass spectra of peptide derivatives is available through detailed scrutiny of the spectrum, for 'metastable peaks'. These appear as broad, low-intensity peaks, usually at non-integral m/z values, and they arise from ions that decompose ($m_1^+ \rightarrow m_2^+$) in the short interval of time ($<10^{-6}$ s) that elapses between ion formation with expulsion from the ion source and arrival at the ion collector in the mass spectrometer. 'Normal' ions are produced by fragmentation processes that are completed in the ion source, so they are recorded as ions of integral m/z. From the m/z value, m^*, of the metastable ion, the ions of masses m_1 and m_2 which also appear in the mass spectrum can be connected together because they have a parent-to-offspring relationship: $m^* = (m_2)^2/m_1$. This relationship can be a useful confirmation of spectral assignments for ion-fragmentation pathways in more complex examples than those so far discussed in this chapter. This is illustrated for pithomycolide, a non-toxic cyclic depsipeptide from *Pithomyces chartarum* (Figure 4.8), containing L-alanine, *N*-methyl-L-alanine, L-2-hydroxy-3-methylbutanoic acid and D-3-hydroxy-3-phenylpropanoic acid in ratios 1:1:1:2. It gives a mass spectrum ($M = 552.2477$) showing small, broad metastable peaks at 237, 400.7, 168.8 and 74.3 atomic mass units. This allows parent ions m_1 to be connected with daughter ions m_2; thus the 332 peak is formed by loss of a neutral grouping (mass $= 465 - 332 = 132$; therefore, —CH(C_6H_5)CH$_2$CO—) from a fragment of mass 465 (since $237 = 332^2/465$). The 316 peak is formed similarly from a fragment of mass 332, likewise for fragments of masses 231 and 131 (Figure 4.8; Rahman *et al.*, 1976).

This interpretation (of the origin of the metastable peaks) eliminates a number of other possible structures and is consistent with the loss of an alanine moiety, then

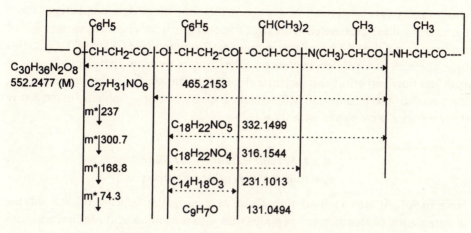

Figure 4.8. Fragmentation of pithomycolide.

phenylpropanoate, then *N*-methylalanine and then L-α-hydroxy-valeric acid (i.e. L-2-hydroxy-3-methylbutanoic acid), from the ionised molecule. Since these losses occur starting from the *C*-terminus of the depsipeptide, the sequence follows as shown in Figure 4.8.

4.8.3 Difficulties and ambiguities

After a substantial range of examples of EIMS mass spectra of peptides had accumulated, it became clear that difficulties and ambiguities of interpretation could be expected (Anderegg *et al.*, 1976). Applications of EIMS methods declined, as 'softer' ionisation methods – such as '*fast-atom bombardment*' (FAB) MS – became established (Biemann and Martin, 1987; Biemann, 1989). These methods create less-energetic molecular ions and the ensuing fragmentation is therefore less extensive (this usually ensures the presence of an intense molecular ion peak in the spectrum) and spectra are therefore more easily interpreted.

4.9 The general status of mass spectrometry in peptide analysis

Why was MS slow to gain favour in peptide structure determination? After all, by 1965, when the early mass spectra of peptides were being collected and assessed, sophisticated examples of structure determination by MS were commonplace in other areas of the organic chemistry of natural products.

Although the interpretation of peptide mass spectra is relatively straightforward, the problem of low volatility – getting enough of the peptide sample into the gas phase to give satisfactory mass spectra – dissuaded many early potential users of electron-impact ionisation MS (EIMS; the only routine technique for mass spectrometry then available). It was about this time that Edman and Begg (1967) demonstrated the automated sequencer ('Sequenator') for chemical sequencing of peptides even if they were available in only small amounts (see Chapter 5). From this time, the idea that peptide-sequence analysis was a problem that could be solved with less time and effort using chemical methods had started to become accepted, but even so, the benefits of MS remained clear as more problems of structure determination were solved with its aid.

4.9.1 Specific advantages of mass spectrometry in peptide sequencing

Nevertheless, it was realised that there were potential benefits of MS for the determination of structures of peptides with unusual amino acid residues and with unknown crosslinking patterns between peptide chains. The importance of cross-links in peptides and proteins (and not only disulphide 'bridges') has been appreciated increasingly; they play a part in the slow deterioration of the organism ('ageing') and their presence in proteins can be linked to certain diseases. One of these, osteoporosis, can be diagnosed through the analysis of protein from a patient's urine sample, from the presence of pyridinoline and deoxypyridinoline in the hydrolysed samples. As illustrated in the preceding section, there are cases of peptides that are difficult or impossible to sequence by standard Edman methodology, but are easily studied by MS.

Some proteins contain asparagine and aspartic acid residues (and glutamic acid and glutamine) and the problem of determining their relative locations is an easier task using MS than using chemical sequencing methods. A common feature in many naturally occurring peptides, particularly oligopeptide hormones, is post-translational C-terminal amidation ($-CONH_2$ instead of $-CO_2H$ at the C-terminus). Although the presence of this C-terminal feature is not difficult to establish by chemical methods of structure assignment, chemical methods can be tedious, whereas MS is well-suited to the task (the $-CO_2H$ and the $-CONH_2$ groups differ by one atomic mass unit).

A substantial proportion of biologically active peptides carry an acylated N-terminus or N-terminal pyroglutamic acid and are therefore not amenable to chemical degradation from the N-terminus (Edman sequence analysis). However, these N-blocked peptides and N-protected peptides, prepared through laboratory peptide synthesis for which one needs routine checks on structure, can be studied by MS.

A further benefit from MS arises in the study of impure peptides, even mixtures of peptides. Thus, a series of mass spectra obtained from a sample as the source temperature is raised can be interpreted to give information about each component in a mixture of peptides.

4.10 Early methodology: peptide derivatisation

The practical problem of getting sufficient quantities of highly involatile compounds into the gas phase was solved for EIMS of short peptides by chemical derivatisation. The upper limit for structure determination of derivatised peptides by MS had been thought to be about twelve amino-acid residues. This was because the lowering in intensity of fragment ions from higher to lower masses that is seen in typical mass spectra causes uncertainties in assignments because the small peaks merge into the background of the mass spectrum. This limitation was not considered to be a drawback since the normal methodology was to establish protein structures through study of overlapping peptides produced by partial hydrolysis. This approach (chemical or enzymatic cleavage into smaller peptides) is still used, even with the advent of FAB MS and other techniques (see Section 4.11) that are capable of generating and analysing ions with m/z values of more than 1000.

Successful examples of structure determination of derivatised components of protein hydrolysates, together with the fact that mass spectra could be obtained with very small samples, provided the impetus for rapid progress towards the current situation, in which a range of mild ionisation techniques and means of dealing with involatile samples are available. Although these improvements extend the range to much higher relative molecular masses without the need either to derivatise or, in favourable cases, to carry out partial hydrolysis, they were not in themselves the breakthrough to wider acceptance of MS in peptide-structure determination; it was the success of peptide derivatisation that generated confidence in mass-spectrometric techniques.

4.10.1 N-Terminal acylation and C-terminal esterification

These procedures were found (Das and Lederer, 1971; Das *et al.*, 1967) to increase the volatility of a peptide by removing some of the intramolecular and intermolecular polar interactions that are characteristic of amino acids, peptides and proteins. The procedures have continuously been developed during more recent years, since they also form the basis of derivatisation for gas–liquid chromatographic (GLC) analysis of amino acids (Section 4.18.2) and peptides.

In fact, some of the early examples of structure determination were particularly favourable in terms of practical MS (even though they appear to be more structurally complex examples than would be expected for pioneering studies) because they were sufficiently volatile without derivatisation. They could be described as 'naturally derivatised' peptides (many natural peptides are *N*-acylated and esterified). Dolastatin 15 (Figure 4.7) has no zwitterionic characteristics, since it is *N*-methylated and the *C*-terminus is cyclised as part of a lactone. It is already suitable for MS since it has adequate volatility without derivatisation.

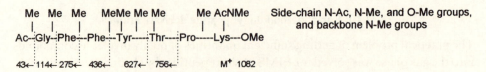

Figure 4.9. Fragmentation of a derivatised heptapeptide.

4.10.2 N-*Acylation and* N-*alkylation of the peptide bond*

One or other of these substitution reactions will further increase the volatility of an *N*-acylated peptide ester, by eliminating hydrogen-bonding or polar interactions between peptide bonds. A procedure once commonly used is Purdie *N*-methylation of all amide nitrogen atoms (accompanied by methylation of all other nucleophilic sites such as —NH_2, —CO_2H, —OH, —SH, —SO_3H, etc.) by reacting the peptide with iodomethane and silver oxide. The same result is sometimes better achieved by adding NaH or $NaCH_2S(O)Me$ to the peptide in DMF, followed by reaction with iodomethane for 5–20 min.

Figure 4.9 shows the mass-spectral details for the heptapeptide H—Gly—Phe—Phe—Tyr—Thr—Pro—Lys—OH, which fails to give an informative EIMS mass spectrum after the simplest derivatisation (*N*-terminal and *N*-Lys-side-chain acetylation and *C*-terminal and Tyr- and Thr-side-chain methylation), but which succumbs when additional methylation of peptide bonds is carried out.

4.10.3 *Reduction of peptides to 'polyamino-polyalcohols'*

An alternative derivatisation procedure involves successive *esterification* (MeOH/HCl), *N*-acetylation (Ac_2O), or N-*trifluoroacetylation* (CF_3CO_2Me), *reduction* with $LiAl^2H_4$ and O-*trimethylsilylation* (with trimethylsilyl chloride). This lengthy procedure gives derivatives (Figure 4.10) that show simple mass spectra owing to C—C bond cleavage along the backbone of the molecule, with the positive charge (cf. Scheme 4.1) located on one side or the other of each cleaved bond. Thus, the mass spectra consist of *two* series of ions, one recorded as the masses of the fragments on the *N*-terminal side of the cleavage points, the other series for *C*-terminal fragment ions. An example (Nau, 1976) is shown in Figure 4.10.

A variety of other derivatisation regimes has been described (Falter, 1971), each with its problems and its advantages. Each has its adherents, but most users of EIMS remain uncommitted to any particular one of them.

4.11 Current methodology: sequencing by partial acid hydrolysis, followed by direct MS analysis of peptide hydrolysates

Treatment of a peptide with 6 M HCl at 100–110 °C for 3–30 min and accurate mass measurement and immonium ion analysis (Section 4.7.1) using plasma desorption

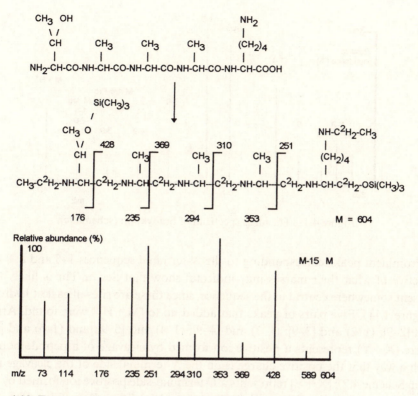

Figure 4.10. Fragmentation and mass spectrum of derivatised H—Thr—(Ala)$_3$—Lys—OH.

MS (see the next section) has been advocated as a simple sequencing protocol (Zubarev *et al.*, 1994; see also Vorm and Roepstorff, 1994). The choice of mass-spectrometric method is guided by the need to produce low-energy ions that do not undergo much fragmentation and therefore can be relied on to give spectra with prominent molecular ions for the hydrolysate components.

Its use with bradykinin and desmopressin (Zubarev *et al.*, 1994), illustrated as a case study below, shows the benefits of MS sequence analysis. The latter peptide, desmopressin ('Minirin', a neurohypophysial hormone analogue), involves a disulphide link and a *C*-terminal amide, as well as a non-coded amino-acid residue (Mpa = mercaptopropionic acid) and these features can be identified by chemical sequencing methods (Chapter 5) only with considerable effort.

Bradykinin   This peptide hormone has a relative molecular mass of 1059.578 ± 0.021 Da for $[M + H]^+$. After 3 min of hydrolysis, the same $[M + H]^+$ ion was present, indicating that Asn, Gln and terminal amide are all absent from the peptide, since ammonia would have been released; and that, since a peak 18 atomic mass units higher was not created by hydrolysis (M plus H_2O), it is not a cyclic peptide.

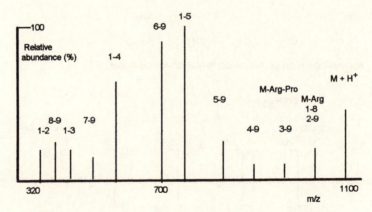

Figure 4.11. The mass spectrum of bradykinin (schematic).

Prominent peaks corresponding to the N-terminal sequences 1–5 and 5–9 (irrespective of what their masses may indicate) show that Ser or Thr is likely to be present somewhere central in the sequence, since these are sites of easiest hydrolysis (Figure 4.11). Five pairs of peaks that added up to $[M + H]^+$ were found: Arg H^+ and [2–9]; (1–2) and [3–9]; (1–3) and [4–9]; (1–4) and [5–9]; and (1–5) and [6–9]. (Here (X − Y) represents a positive ion formed by cleavage of a peptide bond, in such a way that the positive charge is on the N-terminal side of the cleavage point (C in Scheme 4.2); [X − Y] represents a C-terminal-side positive ion formed by fragmentation made 'y' in Scheme 4.4). Therefore Arg—Pro—Pro—Gly—Phe—... is indicated from the mass differences. Matching the differences between [6–9] and [7–9] and [7–9] and [8–9] indicated that the C-terminal tetrapeptide sequence is —Ser—Pro—[8–9].

The peak [8–9] at mass 322.187 ± 0.007 Da is of $[M + H]^+$ type and must be [Arg—Phe]$^+$ or [Phe—Arg]$^+$, although if the experimental error involved were ± 0.012, then [Gly + Phe + Val]$^+$ would remain a possibility. Immonium ion analysis (Section 4.7.1) is in favour of [Arg—Phe]$^+$ (also, no peaks corresponding to losses of Gly, Phe and Val could be found).

The sequence on this basis is H—Arg—Pro—Pro—Gly—Phe—Ser—Pro—Phe—Arg—OH; but the *direction of the sequence could be the reverse of this and remain compatible with all the above arguments*. Since the cleavage of peptides at serine, by partial hydrolysis, is easier for serine residues near the N-terminus than it is for those later in a sequence, comparing the relative intensities of [7–9] and (1–5) (the latter is more prominent) shows that the sequence written above is correct.

4.11.1 Current methodology: instrumental variations

High-resolution MS has considerable merit in sequence analysis of peptides, since the technique allows the atomic composition of parent and fragment ions to be assigned unambiguously through their accurate mass values. However, the sensitivity is an order of magnitude lower, so it is only with short peptides that any benefit would be expected for this reason. Nevertheless, short cuts in the process of assigning structures are provided by accurate mass measurements; it was helpful, for example in assignment of structures to Dolastatin 15 (Figure 4.7), pithomycolide (Figure 4.8) and bradykinin, as described above.

Chemical ionisation mass spectra (CIMS; NH_4^+ as ionisation reactant) can yield satisfactory spectra from underivatised peptides and *field-desorption ionisation* has been shown to give intense $[M + 1]^+$ ions from otherwise involatile peptide derivatives (for example, CH_3—CO—Gly—Arg—Arg—Gly—OCH_3; Buehler *et al.*, 1974), but less sequence information is gained in these mild ionisation procedures because less fragmentation occurs and there are relatively few peaks in the mass spectrum. 'MALDI' – matrix-assisted laser desorption ionisation MS – is an acronym that is encountered in recent literature for this ionisation technique.

^{252}Cf *plasma desorption* (bombardment of the sample with radioactive decay particles in the ion source) gives spectra showing prominent molecular ions (e.g. *m/z* 1904 and *m/z* 1918 from un-derivatised Gramicidins; MacFarlane and Torgerson, 1976) and molecular ions from proteins of relative molecular masses up to 30 000 to within an accuracy of 0.1–0.2%. In ascendancy from the 1980s is *fast-atom bombardment* (FAB), using argon ions (Ar^+) to cause ionisation of the sample. *FAB quadrupole mass spectrometry* of peptides with relative molecular masses up to about 15 000 is now becoming routine using much smaller instruments. This mass range is attainable even when only 2–3 nmol samples are available. Typically, the sample is embedded or dissolved in glycerol to maintain an ion beam from the sample for a sufficiently long time for one to be able to record a spectrum.

A common belief is that FAB mass spectra confirm molecular masses and little else; less fragmentation occurs under FAB ionisation conditions than is needed for sequencing applications, since insufficiently energetic $[M + 1]^+$ ions are formed, but sequence information can be extracted in favourable cases. The energy of these quasi-molecular ions $[M + 1]^+$ can be increased by collisions arranged to occur in the *tandem mass spectrometer* configuration (the so-called MS–MS), whereby ions formed by FAB are selected and led into a second mass spectrometer at 10^{-3} Torr (helium), where they undergo *collision-induced decomposition* (CID) with positive helium ions to give fragment ions. Analysis of these in terms of sequence information is then achievable.

An example of FAB MS arranged to generate sequence information is shown in Figure 4.12 for melittin, a 26-residue peptide amide of known structure from bee venom (Greer, 1989). A molecular ion 28 atomic mass units higher than that of the peptide is present in the mass spectrum, showing that the presumed pure component is accompanied by a component so closely similar that it does not separate when melittin is purified; from its mass and logical reasoning about its origins, it must be the *N*-terminal formyl derivative of melittin. This illustrates one of the

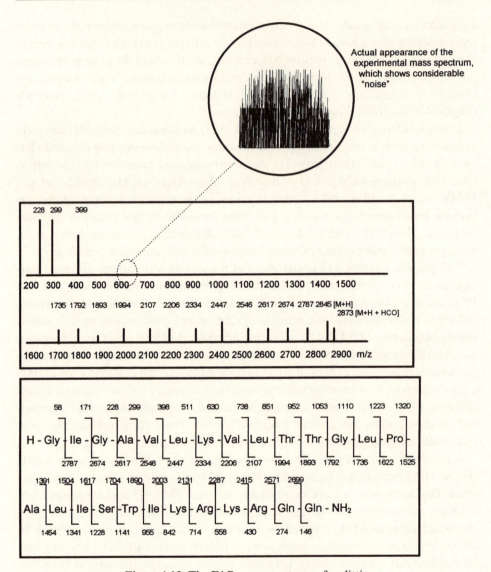

Figure 4.12. The FAB mass spectrum of melittin.

benefits of mass spectrometry in assigning purity criteria when other methods fail, also providing structural information about minor components in mixtures.

In current practice, large peptides are sequenced by FAB MS after tryptic digestion has given fragments up to about 25 amino-acid residues long. The smaller peptides can be sequenced from their mass spectra and unambiguously arranged, jigsaw fashion, to give the structure of the target peptide/protein if a parallel sequencing of fragment peptides obtained using another enzyme with different cleavage proclivities (e.g. the proteinase from *Staphylococcus aureus* strain V8, which cleaves at glutamate residues) is also carried out. A combination of mass-spectrometric structure determination in parallel with Edman sequencing is also one of the current approaches in the protein laboratory, involving confirmation of the structure of the chemically cleaved amino acid residue by various means, such as 'before-and-after' MS of Edman-degraded peptides.

'*Electrospray', 'thermospray' or 'ion-spray' MS* refers to a link of HPLC with MS as a powerful analytical technique, the liquid effluent from the HPLC being sampled and vaporised into the ion source of the mass spectrometer. A powerful electric field acting on the surface of a solution emerging from the HPLC causes nebulisation. The 'mist' of electrically charged droplets is passed to an evaporation chamber in order to evacuate off the solvent and then passed to the ion source of a quadrupole mass spectrometer. $[M + 1]^+$ and $[M - 1]^-$ ions are formed during this process, from small molecules, whereas large peptides yield multiply charged species $[M + 1]^{n+}$ that undergo some fragmentation, whose m/z ratio can be deduced, so providing the relative molecular mass.

4.12 Conclusions

In spite of the power of modern MS, which is capable of analysing large-mass ions with impressive accuracy, problems associated with the generation of suitable ions have to be overcome. An illustration of an unexpected problem is the absolute need to remove inorganic salts from samples subjected to FAB ionisation, otherwise ion generation and expulsion from the glycerol matrix simply does not occur. Many of the problems deriving from molecular structural characteristics of the sample are overcome in ways that have been in use for many years, such as controlled partial hydrolysis of large proteins and derivatisation for enhancing the volatility of samples. The rapidly expanding literature on the subject includes numerous examples demonstrating the individuality of peptides and the ways in which this individuality is reflected in the different conditions needed to give the optimum mass spectrum in each case.

4.13 References

Anderegg, R. J., Biemann, K., Buchi, G. and Cushman, M. (1976) *J. Amer. Chem. Soc.* **98**, 3365.

Biemann, K. and Martin, S. A. (1987) *Mass Spectrom. Rev.*, **6**, 1.

Biemann, K. (1989) in *Protein Sequencing: A Practical Approach*, eds Findlay, J. B. C. and Geisow, M. J., IRL Press, Oxford, p. 99.

Bouchonnet, S., Denhez, J.-P., Hoppilliard, Y. and Mauriac, C. (1992) *Analyt. Chem.*, **64**, 743.

Buehler, R. J., Flanigan, E., Greene, L. J. and Friedman, L. (1974) *Biochemistry*, **13**, 5060.

Compernolle, F., Vanderhaeghe, H. and Janssen, G. (1972) *Org. Mass Spectrom*, **6**, 151.

Das, B. C., Gero, S. D. and Lederer, E., (1967) *Biochem. Biophys. Res. Commun.*, **29**, 211.

Das, B. C. and Lederer, E. (1971) in *New Techniques in Amino Acid, Peptide, and Protein Research*, Eds. Niederweiser, A. and Pataki, G., Ann Arbor Science Publishers, Michigan, p. 175.

Desiderio, D. M. (1991) *Mass Spectra of Peptides*, CRC Press, Boca Raton, Florida.

Edman, P. and Begg, G. (1967) *Eur. J. Biochem.*, **1**, 80.

Falter, H. (1971) in *Advanced Methods in Protein Sequence Determination*, Ed. Needleman, S. B., Springer Verlag, Berlin, Heidelberg, pp. 123–48.

Greer, F. (1989) *Lab. Practice*, October 1989.

Johnstone, R. A. W. and Rose, M. E. (1983) *Mass Spectrometry for Organic Chemists*, Second Edition, Cambridge University Press, Cambridge.

MacFarlane, R. D. and Torgerson, D. F. (1976) *Science*, **191**, 920.

Nau, H. (1976) *Angew. Chem. Int. Ed.*, **15**, 75.

Pettit, G. R. Kamano, Y., Dufresne, C., Cerny, R. C., Herald, C. L. and Schmidt, J. M. (1989) *J. Org. Chem.*, **54**, 6005.

Rahman, R., Taylor, A., Das, B. C. and Verpoorte, J. A. (1976) *Canad. J. Chem.*, **54**, 1360.

Vanderhaege, H. and Parmentier, G. (1971) *Tetrahedron Lett.*, 2687.

Vorm, O. and Roepstorff, P., (1994) *Biol. Mass Spectrom.*, **23**, 734.

Zubarev, R. A., Chivanov, V. D., Hakansson, P. and Sundqvist, B. V. R. (1994) *Rapid Commun. Mass Spectrom.*, **8**, 906.

Part 3. Chromatographic and related methods for the separation of mixtures of amino acids, mixtures of peptides and mixtures of amino acids and peptides

4.14 Separation of amino-acid and peptide mixtures

The two purposes of separation of amino-acid and peptide mixtures are either at the preparative level, to isolate one or more individual components from the mixture for further study; or at the analytical level, to identify and to determine the relative amounts of some or all of the components. Most of the routine studies, conducted daily to determine the amino-acid content of clinical and botanical samples in hundreds of laboratories around the world, are at the analytical level. However, many of the research studies are at the preparative level; an example of this is the identification of crosslinking amino acids from proteins, through their isolation from protein hydrolysates, from physiological specimens for medical investigations, or purely to gain new knowledge.

For peptide mixtures, the objectives are the same as for amino acids. Separation methods provide pure samples for study from natural sources, but analytical methods include the monitoring of peptide synthesis from the points of view both of chemical and of stereochemical purity.

4.14.1 Separation principles

The same principles and instrumentation apply both to preparative and to analytical separation, although a non-destructive means of identifying the separated com-

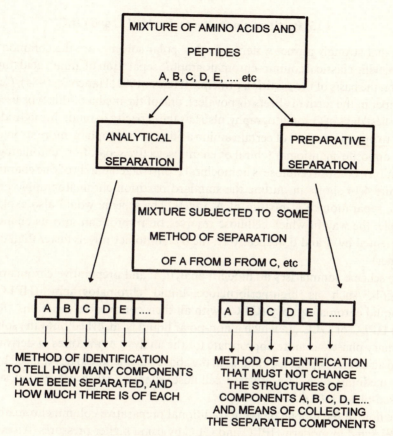

Figure 4.13. Analytical and preparative procedures.

ponents is needed for the latter area. In any preparative separation of a mixture, analytical separation of a mixture is carried out as the initial study to establish separation conditions.

Figure 4.13 illustrates that a separation procedure involves two distinct stages: a means of separation; and a means of identifying the components that have been separated. A means of separation of the components of a mixture can be envisaged for *solid mixtures*, based on the different volatility of each component. Although this is sometimes exploited in mass spectrometry and to some extent in gas–liquid chromatography, all practical procedures in the amino-acid and peptide field are based on separation of components from *solutions*.

In general, *partition* of components from a solution at a solid surface provides the principle that is most often exploited (*adsorption* is used only very rarely), but for amino acids and peptides, which can exist in charged forms in aqueous solutions, *ion-exchange* and *electrophoresis* separation are also available. Separation on the basis of *molecular size* is also used.

4.15 Partition chromatography; HPLC and GLC

Solids that strongly attract water and other polar solvents are the common media for achieving classical column-chromatographic separation of amino acids and peptides, on the basis of the partition principle (Hearn, 1991; Hancock, 1984). Cellulose (i.e. paper in the form of sheets or powder), one of the media of this type used since the earliest days of chromatography, also has the capacity to bind, through adsorbed water, to one enantiomer of certain amino acids, e.g. tryptophan, more strongly than to the opposite enantiomer (chiral or enantioselective separation; chromatographic resolution), because cellulose is homochiral (constructed purely of one enantiomer).

Figure 4.14 shows in outline the standard partition-chromatographic principle for the separation of a mixture of A and B. The scheme would also explain, for example, the way in which cellulose resolves DL-tryptophan into its enantiomers (represented by A and B, respectively; the D enantiomer travels faster than does the L isomer).

The scheme summarises all modern analytical and preparative chromatography protocols, such as high-performance liquid chromatography (HPLC) and gas–liquid chromatography (GLC), with all their conceivable variations. 'Reverse-phase HPLC or GLC', in which a non-polar liquid is adsorbed onto the solid – the stationary phase – is more appropriate for the analysis of mixtures of derivatives of amino acids and peptides. Cellulose in the above scheme would be replaced by a less-polar medium, such as acetylated cellulose, silanised silica gel, etc. in standard reversed-phase HPLC.

The flow of the mobile phase in traditional preparative column chromatography is accelerated in modern HPLC and GLC by using higher pressures. An even flow is ensured by the use of a stationary phase of uniform particle size and durability. Maintenance of constant flow rates and temperatures is also routinely catered for in modern instrumentation. Solvent composition can be precisely varied through the time scale of a separation. For these reasons ensuring uniformity, the term 'high-performance (rather than high-pressure) liquid chromatography', HPLC, has become standard.

The mobile phase for reversed-phase HPLC of peptides often contains additives ('pairing agents') that improve the resolution of components. These are frequently strong acids (octanesulphonic acid or phosphoric acid) and act, together with control of the pH of the aqueous part of the mobile phase, by fine-tuned interactions with the basic and acidic groupings that may be present in the peptides (Figure 4.15).

4.16 Molecular exclusion chromatography (gel chromatography)

The outline in Figure 4.14 also applies to the separation of components of amino-acid and peptide mixtures on the basis of molecular size. In this case, the exclusion

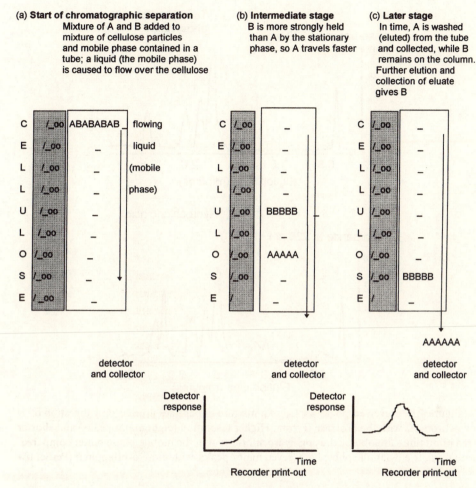

Figure 4.14. '/' and 'oo' represent water and other molecules, adsorbed onto the cellulose (or other chromatographic medium); '/' and 'oo' can also represent functional groups at the surface, to illustrate ion-exchange.

of larger molecules (e.g. A in Figure 4.14, but not B, which has a smaller relative molecular mass) from the pores and crevices of an insoluble polymeric gel means that larger molecules (A) will travel with the mobile phase and therefore emerge first from the column, whereas smaller molecules (B) will be impeded in their flow.

The mechanism of separation may be a little more complicated than this, since partition might also be involved. Standard practice would aim to exclude partition by suitable choice of the mobile phase; however, useful variants of the technique enhance the separation that can be achieved and the gel can be modified by substitution with functional groups. If, for example, these are ionic groups, then ion-

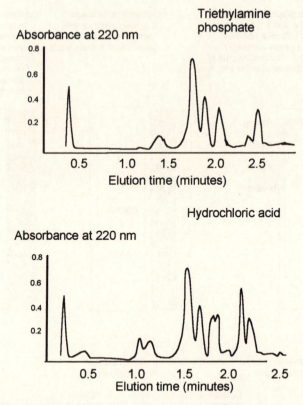

Figure 4.15. Reversed-phase HPLC of a mixture of peptides produced by digestion of cytochrome C with the protease trypsin. Higher resolution (eight major peaks) and shorter retention times are obtained using hydrochloric acid in the mobile phase buffer compared with using triethylamine phosphate (seven major peaks, with more overlapping) (PerSeptive Biosystems Inc.).

exchange separation principles (Section 4.17) can be superimposed on the molecular exclusion principle.

4.17 Electrophoretic separation and ion-exchange chromatography

Because amino acids and peptides in aqueous solutions can receive or donate protons and in the process will gain an overall electrical charge, a stationary phase carrying acidic or basic functional groups will interact differently with the components of a mixture (A and B in the classical separation scheme; Figure 4.14). A fine tuning of the electrical charge carried by amino-acid and peptide molecules will occur in aqueous solutions, providing both positively charged and negatively charged species in ratios determined by the pH. This is the basis of the classical Moore and Stein quantitative amino-acid analysis protocol. This exploits the separation of mixtures through the use of a pH gradient. Figure 4.14 summarises this too; A is a species that, at a given pH, interacts less with the stationary phase than does B and is therefore faster moving with the mobile phase. The detailed explanation of the order of elution of components of an amino-acid and peptide mixture is more subtle than would be expected merely on the basis of the pK_a and pK_b values (Section 3.1) of the solutes, since weakly acidic and weakly basic ion-exchange resins are usually preferred for the purpose, so the equilibrium constants of the interactions of the weakly basic amino group and the weakly acid carboxy group have to play a role. The transfer of protons backwards and forwards between the stationary phase and the solute varies as the solute molecules travel along the column encountering new functional groups with which to set up equilibria.

Modern quantitative amino-acid analysis continues to include electrophoresis and ion-exchange chromatography among the currently available analytical methods, taking advantage of some of the more sophisticated instrumentation developed for HPLC chromatography. Because of the gain of electrical charge in buffers of appropriate pH, movement of amino acids and peptides in a uniform electrical field can be brought about (electrophoresis). The apparatus employed to achieve this uses the stationary phase, over which a constant voltage is maintained, in contact with a stationary or moving liquid (usually an aqueous buffer). The principle is otherwise the same as for the other separation techniques (Figure 4.14). The extra component needed is the means of applying the electric field, together with a means of cooling the system (which is warmed by the passage of current through the buffer) for accurate work. This approach is rarely used for preparative separation of amino acids and peptides; the growing appreciation of the usefulness and flexibility of HPLC is tending to put the use of classical electrophoresis into decline.

4.17.1 Capillary zone electrophoresis (CZE)

Application of some of the instrumentation principles of HPLC to electrophoresis, bearing in mind the need to cause all of the components of a mixture to migrate differently, has led to the development of several related techniques that are particularly useful in the amino-acid and peptide fields (Baker, 1995). A typical electropherogram (Figure 4.16) indicates the salient features of the capillary zone electrophoresis (CZE) analysis protocol.

4.18 Detection of separated amino acids and peptides

Once the components have been eluted, one by one, from the stationary phase, a means of detecting the emergence of each one of them from the column is required. The detection systems applying to liquid–liquid separations (Section 4.18.1) are different from those applying to gas–liquid separations (Section 4.18.2).

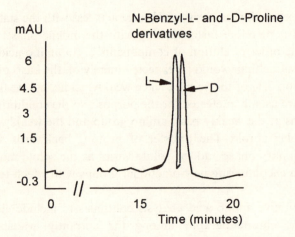

Figure 4.16. Resolution of enantiomers using BioRad Biofocus 3000 capillary electrophoresis; 36 cm × 50 μm capillary column at 20 °C, 100 mM aqueous β-cyclodextrin in pH 2.5 phosphate buffer (Busacca *et al.*, 1996).

4.18.1 Detection of amino acids and peptides separated by HPLC and by other liquid-based techniques

There are few examples nowadays of the classical chromatographic detection method based on visible colour, since the sensitivity is low and other approaches provide the necessary quantitative accuracy. The visible-colour principle, however, remains common in thin-layer chromatographic analysis (TLC; Section 4.19). Elution of amino acids and peptides from liquid-chromatographic columns can be monitored using a short-wavelength light source ($\lambda = 214$ nm commonly; reliable lamps for $\lambda = 200$ nm are becoming more widely used) and UV detectors.

Derivative formation (pre-column derivatisation; Section 4.5) is the most widely used approach that permits detection to be accomplished. It is based on the light-absorption properties of each derivatised amino acid or peptide that emerges from the column. Alternatively, changes in refractive index (RI) of the mobile phase that occur when solutes are present in the eluate can be exploited to detect the arrival of a separated component at the end of a chromatographic column. The RI-measuring detector can be quite simple, or interference-polarising refractometry can be employed, allowing capillary columns to be used, giving detection limits of about 10^{-7} in RI, so that about 10 pg of a polypeptide can be detected in an eluate (Alexander *et al.*, 1992).

This is by no means the only approach, however. The Moore and Stein protocol employs post-column derivatisation with ninhydrin to give a blue colour whose intensity can be measured spectrophotometrically, to provide quantitative detection of each of the separated components.

Greater sensitivity can be attained using fluorescent derivatives (Section 4.5) and estimations of amounts down to femtomole (10^{-15} mol) and (with capillary-zone electrophoresis even attomole (10^{-18} mol) levels are possible. Estimation of enantiomer ratios is conveniently accomplished using diastereoisomer-forming derivatisation protocols and the separation of the diastereoisomers over normal HPLC reverse-phase media, as an alternative to the separation of enantiomer mixtures over chiral stationary phases.

4.18.2 Detection of amino acids and peptides separated by GLC

The detection of samples in the gas phase through katharometry, electron capture, etc. is well developed and reliable. The linking of a gas chromatograph with a mass spectrometer is increasingly found to be useful in research studies, to identify the components of mixtures whose quantitation has been accomplished by other means. Thermospray techniques that have entered into routine use in mass spectrometry (Section 4.11), have made the CZE–MS and various HPLC–MS combinations particularly useful.

Sufficient volatility for GLC analysis is found for N-acylated esters of amino acids and peptides. Their preparation requires a two-step derivatisation protocol and therefore introduces a potential source of error. There is also anxiety about the impurities that may be introduced in this way. However, this applies to any derivatisation protocol and experienced users of the GLC technique can obtain impressive reproducibility of results, sufficient to match the reliability of the classical Moore and Stein procedure. Flexibility because of the additional range of detectors available for GLC can be useful, e.g. highly sensitive electron-capture detectors for halogenated analytes or amino acids and peptides derivatised with halogen-containing groups.

Estimation of enantiomer ratios is conveniently accomplished using diastereoisomer-forming derivatisation protocols or through separations of enantiomer mixtures over chiral stationary phases. Commercially available chiral coatings for this purpose, such as Chirasil-Val, have been used in the field of amino-acid fossil dating (Section 1.11), exploiting the better resolution of capillary GLC, whereby a thermally stable liquid coats the surface of a narrow tube.

A representative example of GLC support to peptide synthesis concerns alamethicin F-30, a peptide antibiotic from *Trichoderma viride* (Figure 4.17; Akaji *et al.*, 1995). The natural product was synthesised through a classical stepwise solution method and verification of its structure and stereochemical integrity was achieved by various analytical methods. Studied as its sodium salt, the product of synthesis had the RMM 1986.104 = $[M + Na]^+$ by high-resolution FAB mass spectrometry (Section 4.11.1), corresponding to the molecular formula $C_{92}H_{150}N_{22}O_{25}Na$ (which has RMM = 1986.101). The amino acids it contains, which were detected through hydrolysis and derivatisation as their N-pentafluoropropionyl n-butyl esters and capillary GLC analysis over Chirasil-L-Val, were found to be free from D enantiom-

Ac-Aib-Pro-Aib-Ala-Aib-Ala-Gln-Aib-Val-Aib-Gly-Leu-Aib-Pro-Val-Aib-Aib-Glu-Gln-Pheol

Figure 4.17. Alamethicin F-30 (Pheol is phenylalaninol, i.e. phenylalanine with its carboxy group reduced to CH_2OH).

ers above a level of 0.5% (a level that these authors claim is introduced into amino acids when a peptide is hydrolysed).

4.19 Thin-layer chromatography (planar chromatography; HPTLC)

The simplest technique that can accomplish the separation of amino-acid and peptide mixtures, or mixtures of derivatised amino acids and peptides, involves a paper sheet (paper chromatography) or a thin layer of adsorbent immobilised on a glass plate (thin-layer chromatography; TLC) as the stationary phase. Figure 4.14 also applies to explain the principles of TLC methods; the mobile phase travels through the stationary phase on the basis of capillary action in the most commonly used version (ascending chromatography). The detection methods are fairly primitive since they are usually based on visual comparisons after spraying with a colour-forming reagent or after UV irradiation. Semi-quantitative analysis (i.e. obtaining rough numerical data through comparisons of spot areas) can be achieved.

The techniques have their uses for rapid and simple monitoring of mixtures to determine the approximate relative amounts of components. Preparative TLC is often useful to purify the product of a small-scale synthesis (e.g. 0.25 mm silica gel layers and elution of peptides with a 6:3:1 mixture of EtOAc:MeOH:water as the mobile phase, to isolate 4–10 mg of a peptide product). Attempts to make the method more sophisticated, to give reliable quantitative information, have been largely unsuccessful. Perhaps the simplification and wide availability of HPLC techniques have suppressed interest in furthering the role of TLC for analysis of mixtures of amino acids, but improved stationary phases have contributed to better reproducibility (HPTLC), and routine TLC monitoring to validate the purity of intermediates in peptide syntheses is widely used (Barlos *et al.*, 1993).

TLC using silica gel coated with an L-prolinamide–copper(II) salt mixture (Chiralplates; Macherey-Nagel Co.) separates enantiomers using the ligand-exchange principle to give information on the chiral purity of amino acids and peptides. The equivalent HPLC procedure has been used for determining enantiomer ratios.

4.20 Quantitative amino-acid analysis

HPLC and GLC of derivatised amino acids are overtaking the classical 'amino acid analyser' (the Moore and Stein ion-exchange separation plus post-column ninhydrin system) in routine laboratory use for quantitative amino-acid analysis. There is, on this basis, some competition among the various derivatisation protocols (choice of derivatisation and of separation method) and it is clear that each of the methods, when practised by laboratory staff, is capable of delivering reliable results. An additional benefit of HPLC and GLC methods is their capacity for enantiomeric analysis, a facility lacking in the Moore and Stein system.

The N-*phenylthiocarbamoylation (PTC) protocol* is the basis of a commercial system (the Waters Pico-Tag system) that embodies a thoroughly worked-out procedure using semi-automatic equipment with the aim of ensuring that every mixture analysed is treated in an identical fashion. However, many more examples of systems, custom-built around a favoured HPLC or GLC, are featured in the current literature, particularly using the OPA/fluorescence (Section 4.5.1) and *N*-Fmoc derivatisation methods, rather than the PTC-derivatisation method.

The importance of slavish adherence to all details of a protocol in order to obtain reliable results is emphasised by all analysts in this field. This applies even more stringently to the sample-preparation and derivatisation stages, including the preliminary stages in GLC analyses for amino acid determinations.

4.21 References

Akaji, K., Tamai, Y. and Kiso, Y. (1995) *Tetrahedron Lett.*, **36**, 9341.
Alexander, M. L., Belenkii, B. G., Gotlib, V. A. and Kever, J. E. (1992) *J. Microcolumn Sep.*, **4**, 385.
Baker, D. R. (1995) *Capillary Electrophoresis*, Wiley, New York.
Barlos, K., Gatos, D., Papaphotiou, G. and Schafer, W. (1993) *Liebigs Ann. Chem.*, 215.
Busacca, C. A., Dong, Y. and Spinelli, E. M. (1996) *Tetrahedron Lett.*, **37**, 2935.
Hearn, M. T. W. (1991) *HPLC of Proteins, Peptides, and Polynucleotides*, VCH Publishers, New York.
Hancock, W. S. (Ed.) (1984) *Handbook of HPLC for the Separation of Amino Acids, Peptides, and Proteins* (Volumes 1 and 2), CRC Press, Boca Raton, Florida.

Part 4. Immunoassays for peptides

4.22 Radioimmunoassays

If a peptide contains Tyr, it is possible to iodinate this amino acid residue with ^{125}I in the *ortho* position relative to the hydroxy group. This is effected by reaction of the peptide with Na[125]I in the presence of chloramine-T. If a known amount of the labelled peptide (P*) is allowed to compete with a measured volume of a solution containing an unknown concentration of unlabelled peptide (P) for a known limited amount of antibody (Ab) raised to the unlabelled peptide there will be a competition for the antibody binding sites:

$$P + Ab \rightleftarrows P{-}Ab$$
$$P^* + Ab \rightleftarrows P^*{-}Ab.$$

If the antibody is immobilised on 'Sepharose', the supernatant containing the free, radioactive peptide can be separated easily and assayed in a gamma counter. With a standard curve drawn for known amounts of peptide subjected to assay under exactly the same conditions, unknown amounts of peptide can be determined by interpolation on the standard curve. There are two potential problems with this type of radioimmunoassay. First, the peptide to be assayed perhaps does not contain Tyr. If it contains His, however, this may suffice since His can be iodinated, especially by an enzymic procedure described below. Alternatively, the peptide is allowed to react with the Bolton and Hunter reagent (Bolton and Hunter, 1973), prepared by iodination of the ester of 3-(4'-hydroxyphenyl)propionic acid and *N*-hydroxysuccinimide. Any free amino group can be acylated by this reagent. Secondly, reaction of a peptide with NaI and chloramine-T can cause oxidation of Met, Cys and even Tyr residues, which can interfere with complexation of the iodinated peptide with antibodies raised to the un-iodinated peptide. An alternative method (Holohan *et al.*, 1973) of iodination uses lactoperoxidase in the presence of H_2O_2. As pointed out above, this procedure is applicable to the iodination of His residues. This method avoids modification of the side-chains of Met, Cys and Tyr.

4.23 Enzyme-linked immunosorbent assays (ELISAs)

Although a great many radioimmunoassays have been established for peptides and proteins, there has been a move in recent years to develop the alternative technique of using enzyme-linked immunoassays. There are several advantages to be gained. The capital cost associated with the use of radioisotopes is considerable (e.g. the need for an approved laboratory design and the cost of gamma counters). Running costs for purchase of radioactive NaI are considerable. As indicated above, iodination of the peptide to be assayed can lead to chemical modification, which in turn may decrease the affinity for its antibody. No modification of the peptide to be assayed is required when using ELISAs.

The experimental protocol for ELISAs (Wisdom, 1994) is depicted in Figure 4.18. The assays are usually carried out using a polystyrene plate with wells to hold reactants. A solution of antibodies against the peptide to be assayed is applied to the wells. The proteins stick to the polystyrene surface by hydrophobic bonding. An aliquot of solution containing the peptide to be assayed is then added under conditions such that the peptide will be bound by the antibody. Any unbound peptide is washed away. This second antibody to the protein to be assayed is added. This second antibody should be specific for binding a different part of the peptide molecule so that the first antibody does not interfere with the binding of the second. The second antibody must have a suitable enzyme covalently attached to it and the conjugated

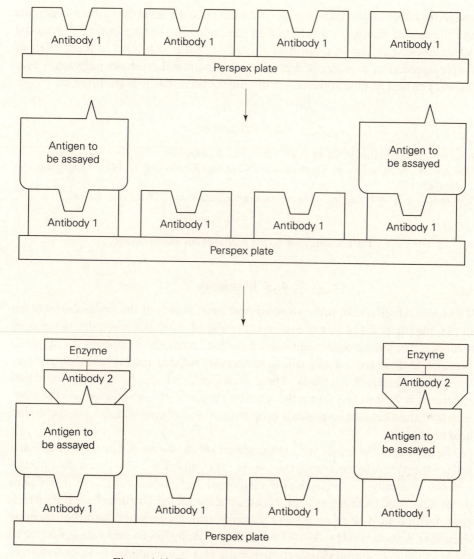

Figure 4.18. Enzyme-linked immunosorbent assay.

enzyme should have its active site accessible to the substrate. This enzyme attached to the second antibody is the signalling device which quantifies the amount of the second antibody bound to the substrate being assayed. It should operate on a substrate that produces a substantial change in absorption of light or in fluorescence. A phosphatase with either 4-nitrophenyl or 4-methylumbelliferyl phosphate is a suitable system. Obviously, the pH of the enzyme–substrate system should not impede the binding of the antibody to the peptide being assayed. Again, the signalling enzyme should be free from proteinases that could degrade either the peptide being assayed or either of the antibodies. A calibration curve linking the amount of peptide and the rate of liberation of product by the signalling enzyme must be produced before assaying unknown quantities of peptide. With both types of immunoassay, simple injection of a conjugated peptide into an animal produces polyclonal antibodies. The best results, however, are obtained with monoclonal antibodies.

4.24 References

Bolton, A. E. and Hunter, W. M. (1973) *Biochem. J.*, **133**, 529.

Holohan, K. N., Murphy, R. F., Buchanan, K. D. and Elmore, D. T. (1973) *Clin. Chim. Acta*, **45**, 153.

Wisdom, G. B. (1994) *Peptide Antigens: A Practical Approach*, IRL Press, Oxford.

Part 5. Enzyme-based methods for amino acids

4.25 Biosensors

This topic is included in order to contribute brief details of the principles involved in the quantitative analysis of amino acids and of some oligopeptides in aqueous solutions. The simple equipment needed so that an electrical response owing to the presence of these compounds will be generated and measured is based on particular enzyme-catalysed reactions. These create net changes of electrical potential generated in various ways from the reaction products that are created and the magnitude of the electrical response is proportional to the concentration of the amino acid or peptide.

The essence of the sensor is an electrode on which an enzyme, or a whole cell that utilises the particular amino acid or peptide, is immobilised. The electrode is part of a circuit generally similar to the pH meter and the whole is calibrated using standards. Glutamate-sensing systems based on glutamic acid oxidase have been developed for this purpose of estimating glutamic acid, aspartic acid and the dipeptide derivative aspartame (H—Asp—Phe—OMe). The hydrogen peroxide generated in the reaction is quantitated amperometrically at a platinum electrode (Suleiman *et al.*, 1992). Glutamate dehydrogenase acts on glutamic acid to generate chemiluminescence when appropriate reagents, including luminol, are added to the sample (Girotti *et al.*, 1992). The intensity of light emitted is proportional to the concentration of glutamic acid in the sample, without interference from other amino acids.

4.26 References

Girotti, S., Ghini, S., Budino, R., Pistillo, A., Carrea, G., Bovara, R., Piazzi, S., Meroighi, R. and Roda, A. (1992) *Analyt. Lett.*, **25**, 637.

Suleiman, A. A., Villarta, R. L. and Guilbault, G. G. (1992) *Bull. Electrochem.*, **8**, 189.

5

Determination of the primary structures of peptides and proteins

5.1 Introduction

The structure of a protein can be considered at four levels. The *primary* structure comprising the sequence of amino acids in the chain(s) is the subject of this chapter. Secondary, tertiary and quaternary structures are described in Chapter 2.

Although the determination of the primary structure of insulin by Sanger in the early 1950s evoked great excitement and earned him the first of two Nobel prizes, some of this chapter is largely of historical interest since Sanger earned himself a second Nobel price by developing a rapid method for sequencing the DNA that codes for proteins. Only twenty amino acids are coded for by DNA (see Chapter 8); related amino acids may arise in peptides and proteins by post-translational modification. Consequently, determination of the primary structure from the DNA sequence does not provide information about post-translational modification and these details must be determined by the classical methods of amino-acid sequencing described in this chapter. Emil Fischer's suggestion at the beginning of the twentieth century that proteins are composed of amino acids linked through peptide bonds (—CONH—), in which the —CO— and —NH— moieties originate from the carboxy and amino groups of consecutive amino acids, has been fully vindicated by synthetic, degradative and X-ray crystallographic techniques. Other covalent bonds link amino-acid residues in peptides and proteins. The commonest is the disulphide bond of cystine, which is formed by oxidation of the thiol groups of two cysteine residues. The disulphide bond may form either a loop within a peptide chain or a crosslink between two separate chains (**5.1**). It should be noted that intermolecular disulphide bonds arise by proteolytic excision of a peptide from a precursor containing one or more intramolecular disulphide bonds. This apparently devious biosynthetic route allows the protein chain of the proprotein to fold so that the members of the correct pair of cysteine residues are adjacent to form the required disulphide bond by oxidation. The residues subsequently excised serve as a molec-

91

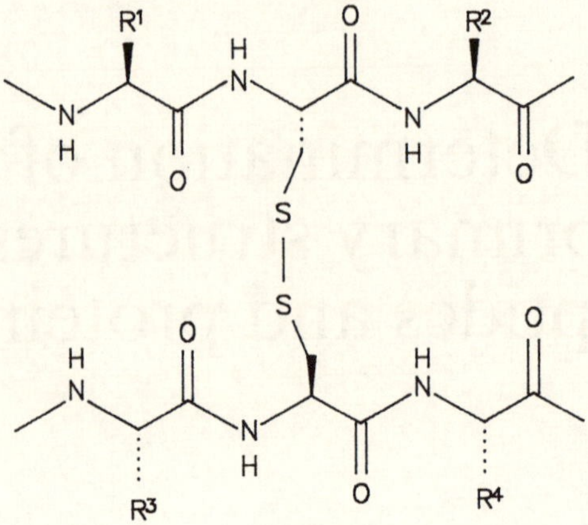

(5·1)

ular jig and force the adoption of a favourable conformation for this process. Thus, pro-insulin comprises the B chain of insulin followed successively by thirty-five amino acids known as the C or connecting peptide and then the A chain of insulin. The three disulphide bonds (A^6—A^{11}, A^7—B^7 and A^{20}—B^{19}) in pro-insulin can be reduced (Section 5.3) and subsequently reoxidised in air to give a high yield of pro-insulin. In contrast, reduction of the intramolecular (A^6—A^{11}) and intermolecular (A^7—B^7 and A^{20}—B^{19}) disulphide bonds of insulin followed by reoxidation in air gives a very low recovery of hormone. The synthesis of the extra peptide bonds in the C peptide is the biological and entropic price required to ensure efficient production of the active hormone.

Several other types of covalent crosslinks, mostly derived from lysine or 5-hydroxylysine residues (the latter being formed by post-translational modification), are found in collagen and elastin. A few examples are given (5.2–5.7): $\Delta^{6,7}$-dehydrolysinonorleucine (5.2), lysinonorleucine (5.3), dehydrohydroxylysinonorleucine (5.4), lysino-5-ketonorleucine (5.5), desmosine (5.6) and isodesmosine (5.7). An intrachain thiol ester loop is present in α_2-macroglobulin and proteins of the complement system and consists of a fifteen-membered ring derived from cysteine and glutamic acid (5.8).

5.2 Strategy

The general strategy for determining the sequence of amino acids in a peptide or protein involves several steps. It is necessary (a) to hydrolyse the molecule completely and to determine quantitatively the relative molar proportions of amino acids present (Chapter 4); (b) to determine the molecular weight in order to calculate the number of residues of each amino acid present; (c) to determine how many peptide chains are present and to separate these, bearing in mind that these may be

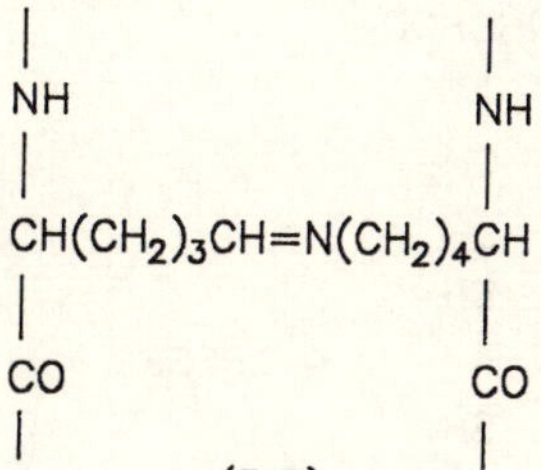

(5·2)

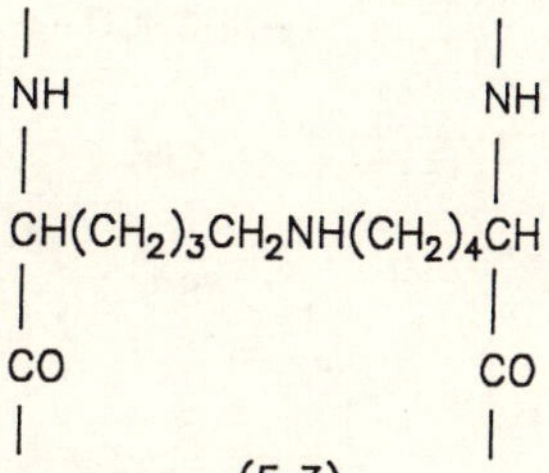

(5·3)

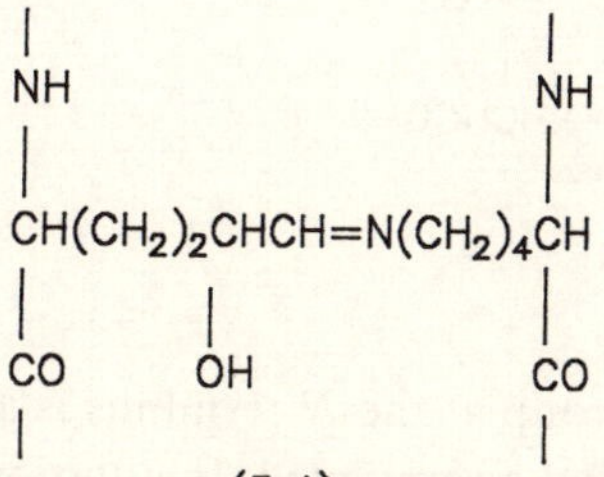

(5·4)

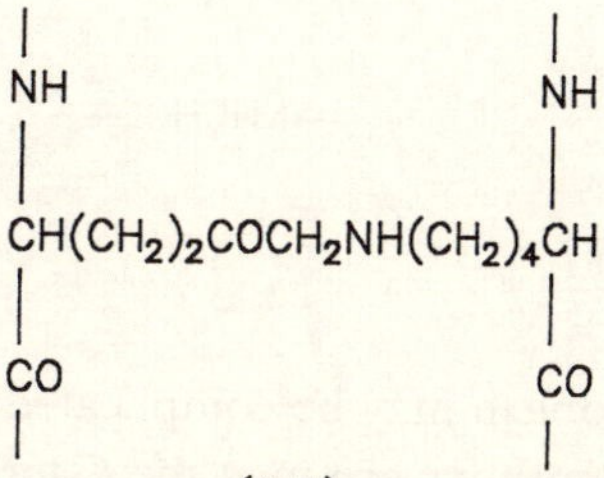

(5·5)

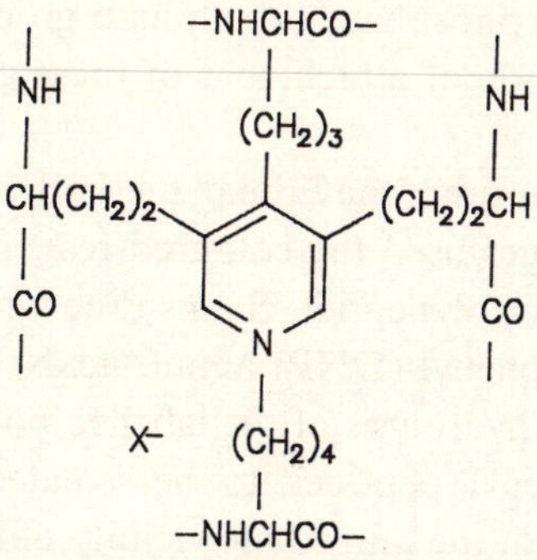

(5·6)

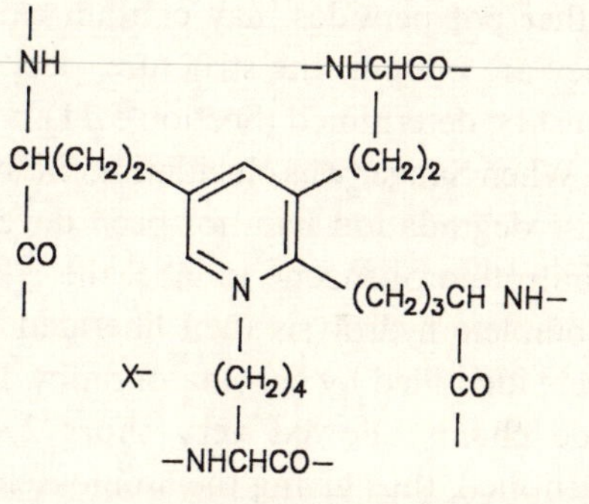

(5·7)

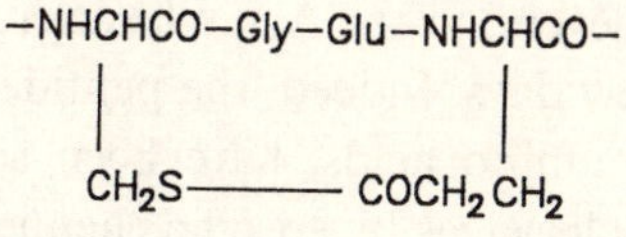

(5·8)

covalently linked, *inter alia*, through disulphide bonds (Section 5.3; Scheme 5.1); and (d) to cleave each peptide chain by specific methods into fragments of convenient size (Sections 5.8 and 5.9) for sequencing by, for example, the Edman method of stepwise degradation (Section 5.4). It should be noted that, under step (c), if the number of chains is determined by analysis of *N*-terminal or *C*-terminal residues,

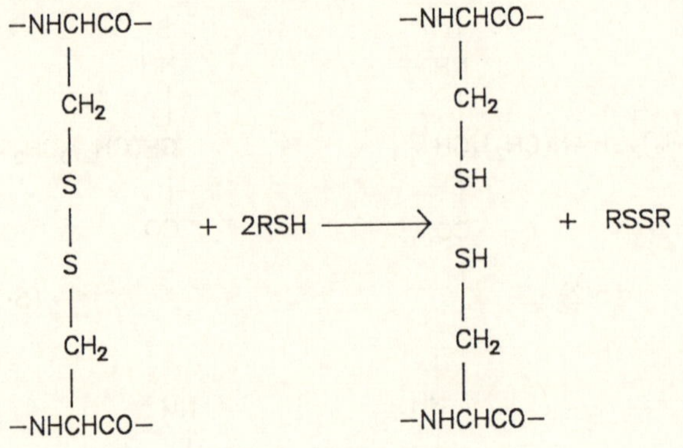

$$-NHCHCO- \qquad\qquad -NHCHCO-$$
$$| \qquad\qquad\qquad\qquad |$$
$$CH_2 \qquad\qquad\qquad CH_2$$
$$| \qquad\qquad\qquad\qquad |$$
$$S \qquad\qquad\qquad\qquad SH$$
$$| \qquad + \ 2RSH \ \longrightarrow \qquad\qquad + \ RSSR$$
$$S \qquad\qquad\qquad\qquad SH$$
$$| \qquad\qquad\qquad\qquad |$$
$$CH_2 \qquad\qquad\qquad CH_2$$
$$| \qquad\qquad\qquad\qquad |$$
$$-NHCHCO- \qquad\qquad -NHCHCO-$$

Scheme 5.1.

the problem may be complicated if the amino group at the *N*-terminus is acylated or the carboxy group at the *C*-terminus is esterified or amidated. In addition, some enzymes contain covalently bound groups (prosthetic groups) at the active site. Other polypeptides may contain carbohydrate, phosphate or sulphate groups. If they are present, the structure, position and mode of attachment of these groups must be determined (Section 5.11).

When Sanger was elucidating the structure of insulin, the Edman method of step-wise degradation had not been developed. Sanger used the coloured reagent 2,4-dinitrofluorobenzene to label the *N*-terminus of polypeptide chains (Scheme 5.2). Complete hydrolysis then liberated the dinitrophenyl (DNP) amino acids, which were identified by chromatography. Partial acid hydrolysis of the labelled polypep-tide chains allowed very short 2,4-dinitrophenyl peptides to be isolated and identified, thus giving the amino-acid sequence in the immediate vicinity of the *N*-terminus. Restricted cleavage of the polypeptide chains using proteolytic enzymes gave other fragments that were similarly identified. Although insulin contains only fifty-one amino acids in two chains, the experimental work involved in determining its structure extended over several years. A similar exercise with modern techniques would be completed in a few days. Indeed, the peptide chains of collagen, each of which contains over 1000 amino acids, have been sequenced. Sequencing of a protein nowadays involves cleavages by specific chemical methods (Section 5.8) or proteinases (Section 5.9) into relatively large fragments. Overlapping fragments can be obtained by using different methods of specific cleavage. In a second revolution in protein structure determination, it has become common to determine the sequence of the DNA which codes for the protein rather than that of the protein itself. The development of the necessary methodology earned Sanger a second Nobel prize in 1980. One of the consequences of this technological revolution is the determination of the sequence of a protein that has never been isolated. In spite of these exciting discoveries, classical methodology has to be used (i) to check the

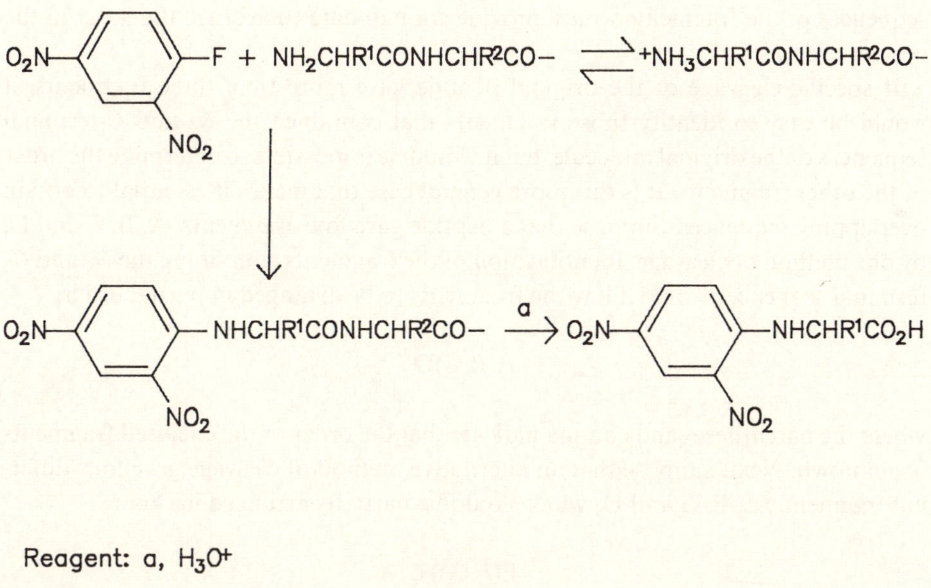

O₂N—[ring]—F + NH₂CHR¹CONHCHR²CO— ⇌ ⁺NH₃CHR¹CONHCHR²CO—

NO₂

↓

O₂N—[ring]—NHCHR¹CONHCHR²CO— —ᵃ→ O₂N—[ring]—NHCHR¹CO₂H

NO₂   NO₂

Reagent: a, H₃O⁺

Scheme 5.2.

sequence of a synthetic polypeptide, (ii) to identify and locate amino acids in naturally occurring peptides and proteins that have undergone post-translational modification and (iii) to determine the position and mode of linkage of prosthetic groups in naturally occurring proteins, since the last two problems are not solved by determination of the nucleotide sequence of the gene.

The importance of the strategy of sequencing using overlapping peptides can be illustrated by some examples. Suppose that a peptide has Ala at the *N*-terminus and Asp at the *C*-terminus and that specific cleavage by some means produces three fragments, A, B and C:

Ala . . . Arg   Val . . . Lys   Leu . . . Asp

A   B   C

(The *N*-terminus has an α-amino group that is not coupled to the carboxy group of another amino acid, although it may be blocked by an acetyl group, for example. The *C*-terminus has an α-carboxy group that is not coupled to the amino group of another amino acid, although it may be present as a primary amide). Since only A contains *N*-terminal Ala and only C contains *C*-terminal Asp, the order of the fragments in the original peptide is A—B—C. The problem would have been only slightly more complicated had two of the fragments had the same *N*-terminal or *C*-terminal residue as the original molecule. Determination of a short sequence of amino acids at either end of the peptide and comparison of the corresponding sequences of the fragments would provide enough data to arrange the latter in the correct order.

If specific cleavage of the original peptide gave more than three fragments, it would be easy to identify those fragments that contained the *N*- and *C*-terminal sequences of the original molecule, but it would be impossible to determine the order of the other fragments. It is this more general case that makes it essential to obtain overlapping sequences. Suppose that a peptide gave four fragments, A, B, C and D, by one method of cleavage. Identification of the fragments containing the *N*- and *C*-terminal sequences would allow the fragments to be arranged in partial order:

$$A(B,C)D$$

where the parentheses and comma indicate that the order of the enclosed fragments is unknown. Next, suppose that an alternative method of cleavage gave four different fragments, E, F, G and H, which could be partially arranged in order:

$$E(F,G)H.$$

If E is longer than A, it will contain towards its *C*-terminus part of the *N*-terminal sequence of B or C. Conversely, if E is shorter than A, part of the sequence towards the *C*-terminus of A will appear as the *N*-terminal sequence of F or G. The reader should now be able to deduce (i) how to achieve the same end by comparing the sequences of D and H; and (ii) how to determine the order of fragments in a peptide when degradation yields more than four fragments.

As indicated above, some amino acids may undergo post-translational modification. In some cases, this may assist one in sequencing the peptide, since the structural modification acts as a built-in label. For example, phosphorylation of the side-chain of a serine residue enables one to keep track of isolated fragments containing phosphoserine and to identify overlapping sequences in its vicinity. In other cases, post-translational modification can lead to complications. The commonest example involves the oxidation of two cysteine residues to form a disulphide bond (Sections 5.3 and 5.10). In glycoproteins, carbohydrate residues are usually attached to the side-chains of asparagine, serine or threonine and may interfere with cleavage by proteinases during the acquisition of fragments for sequence determination. More seriously, the number of pentose or hexose units attached at a particular point may vary from molecule to molecule, giving rise to several fragments from the same part of the molecule.

5.3 Cleavage of disulphide bonds

As indicated in Section 5.2, disulphide bonds may be either intra- or inter-chain ones and end-group analysis might permit differentiation between these possibil-

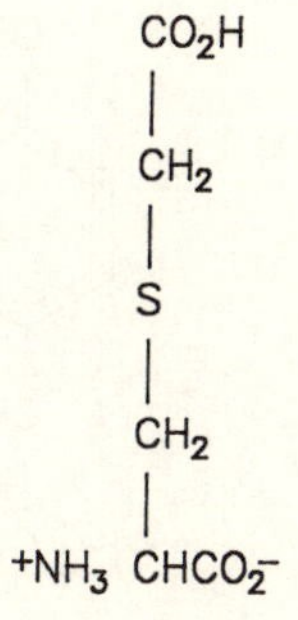

$$
\begin{array}{l}
\text{CO}_2\text{H} \\
| \\
\text{CH}_2 \\
| \\
\text{S} \\
| \\
\text{CH}_2 \\
| \\
{}^+\text{NH}_3\ \ \text{CHCO}_2^-
\end{array}
$$

(5·9)

$$
\begin{array}{ll}
\text{SO}_3^- & \text{H}^+ \\
| \\
\text{CH}_2 \\
| \\
{}^+\text{NH}_3\ \ \text{CHCO}_2^-
\end{array}
$$

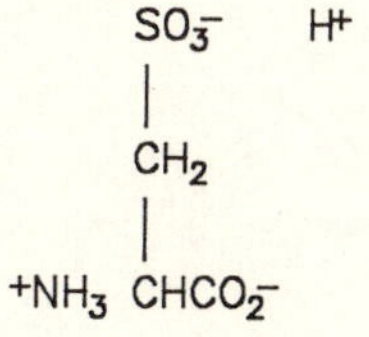

(5·10)

ities, although both may be present in a molecule. Disulphide bonds must be cleaved and, if they link two chains, the latter must be separated and their sequences separately determined. Ultimately, the disulphide bonds must be located in the original molecule. This problem is dealt with later (Section 5.8).

It is usual to effect cleavage of disulphide bonds by reduction or oxidation. Addition of a large excess of a thiol such as 2-mercaptoethanol or 1,4-dithiothreitol to a polypeptide reduces cystine residues to cysteine (Scheme 5.1). In order to prevent reoxidation in air, the generated thiol groups are blocked, usually by reaction with iodoacetic acid. The product yields *S*-carboxymethylcysteine (**5.9**) on hydrolysis for amino-acid analysis. Alternatively, oxidative cleavage of disulphide bonds can be achieved with performic acid; each half of the cysteine residue is converted into a residue of cysteic acid (**5.10**).

5.4 Identification of the *N*-terminus and stepwise degradation

The Sanger method for labelling and identifying the *N*-terminal amino acid has been mentioned above and, following the advent of the Edman method for stepwise degradation, little effort has been deployed in improving the earlier method. One technique worth mentioning is the use of 1-dimethylaminonaphthalene-5-sulphonyl ('dansyl') chloride (**5.11**) in place of 2,4-dinitrophenyl-4-fluorobenzene. Dansylamino acids are strongly fluorescent and so detection and spectrofluorometric assays are much more sensitive. As little as 100 pmol of a dansylamino acid is sufficient for detection and identification by TLC. It should be noted that the sidechains of certain amino acids (**5.12–5.15**) are likely to be labelled and, if these amino acids are *N*-terminal, they can be doubly labelled (e.g. **5.16**). There is no confusion in identifying labelled *N*-terminal amino acids, since only these are extracted into solvents such as ethyl acetate.

Edman's method of stepwise degradation (Edman, 1949, 1950) (Scheme 5.3) involves reaction of the α-amino group at the *N*-terminus of a peptide with phenyl isothiocyanate under slightly basic conditions. Excess reagent is extracted and the

97

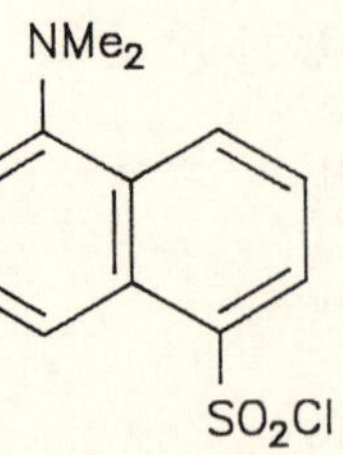

NMe$_2$

SO$_2$Cl

(5·11)

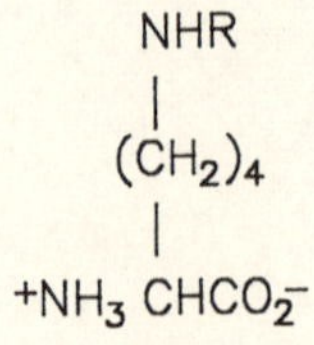

NHR
|
(CH$_2$)$_4$
|
$^+$NH$_3$ CHCO$_2^-$

(5·12)

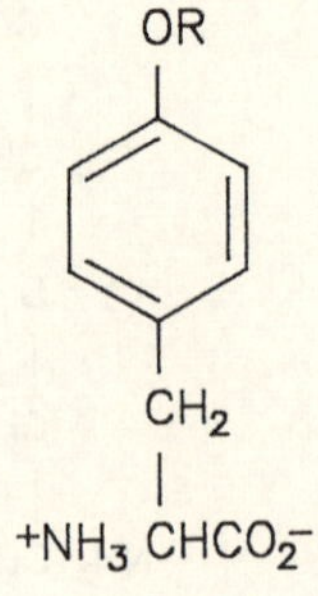

OR

CH$_2$
|
$^+$NH$_3$ CHCO$_2^-$

(5·13)

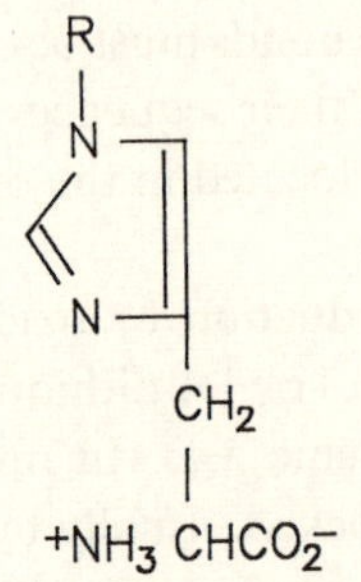

R
N
N
CH$_2$
|
$^+$NH$_3$ CHCO$_2^-$

(5·14)

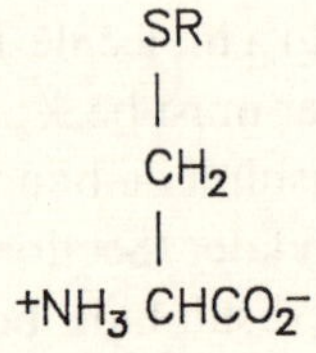

SR
|
CH$_2$
|
$^+$NH$_3$ CHCO$_2^-$

(5·15)

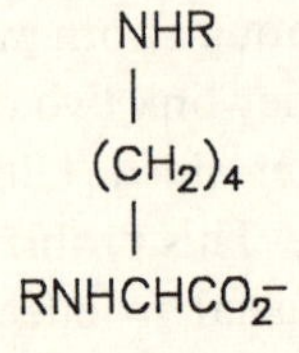

NHR
|
(CH$_2$)$_4$
|
RNHCHCO$_2^-$

(5·16)

$$R = O_2N\text{—}\underset{NO_2}{\bigcirc}\text{—} \quad or \quad \underset{SO_2-}{\overset{NMe_2}{\bigcirc\bigcirc}}$$

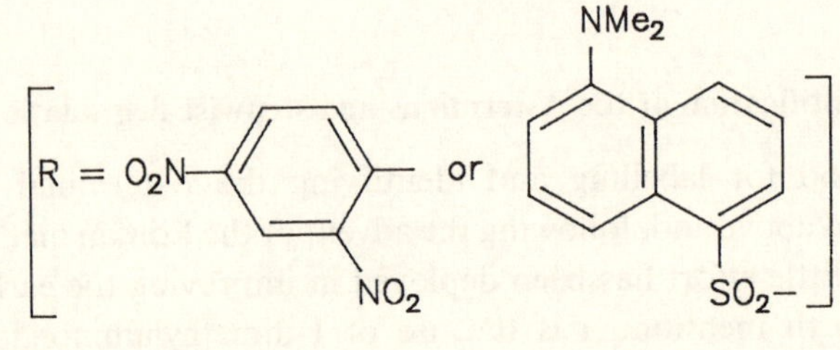

resultant α-phenylthiocarbamoyl peptide (**5.17**) undergoes cyclisation and degradation at the first peptide bond on treatment with a strong acid such as trifluoroacetic acid (Elmore, 1961). This reaction involves nucleophilic attack by the sulphur atom on the carbonyl carbon atom of the peptide bond to give a 2-anilinothiazolin-5-one (**5.18**). This undergoes rearrangement in hot trifluoroacetic acid, probably by ring-opening to the *N*-phenylthiocarbamoylamino acid (**5.19**) and subsequent ring-closure, to give the 3-phenyl-2-thiohydantoin (**5.20**).

The 3-phenyl-2-thiohydantoins can be separated and identified by TLC on silica gel. If the latter contains a fluorophore, the thiohydantoins quench the fluorescence

$$C_6H_5N{=}C{=}S \quad + \quad NH_2CHR^1CONHCHR^2CO{-} \quad \rightleftharpoons \quad {}^+NH_3CHR^1CONHCHR^2CO{-}$$

$$C_6H_5NHCSNHCHR^1CONHCHR^2CO{-} \quad \xrightarrow{a} \quad C_6H_5N{=}C\begin{array}{c} S{-}CO \\ | \\ N{-}CHR^1 \\ H \end{array} \quad + \quad {}^+NH_3CHR^2CO{-}$$

(5·17)

(5·18)

$$C_6H_5N\begin{array}{c} S \\ \| \\ C{-}NH \\ | \\ C{-}CHR^1 \\ \| \\ O \end{array} \quad \xleftarrow{b} \quad C_6H_5NHCSNHCHR^1CO_2H$$

(5·19)

(5·20)

Reagents: a, CF$_3$CO$_2$H; b, hot aqueous acid

Scheme 5.3.

and can be detected as dark spots in ultraviolet light (254 nm). If [^{35}S]-phenyl iso-thiocyanate is used for the Edman procedure, the method is both highly sensitive and quantitative. It is now more usual to identify and quantify the 2-anilinothia-zolin-5-ones or the 3-phenyl-2-thiohydantoins by reversed-phase HPLC on one column using an ultraviolet-absorption detector. This methodology is also used for amino-acid analysis (Chapter 3).

Obviously, incomplete reaction and losses during manipulation prevent the yield of 3-phenyl-2-thiohydantoin from reaching 100%. With each cycle of the Edman method, the yield of product derived from the newly exposed *N*-terminus will decrease. In addition, small amounts of the 3-phenyl-2-thiohydantoins correspond-ing to earlier positions in the sequence will be formed as a consequence of incom-plete reaction at each cycle. If the fraction of peptide that reacts with phenyl isothiocyanate and gives the relevant 3-phenyl-2-thiohydantoin is x, then the yield at any stage is

99

$$x^m(1-x)^{n-m}$$

where m is the position of the particular amino acid in the sequence and n is the number of cycles. Depending on the value of x, a time will come when the yield of 3-phenyl-2-thiohydantoin from the newly exposed N-terminus will be so low and the number of thiohydantoins resulting from incompletely degraded peptide chains will be so large that the chromatograms will be uninterpretable. You will find it instructive to calculate the yields of products from the above formula using various values of x, m and n.

The best results are obtained by using precisely standardised conditions and continuing the degradation without interruption. Edman designed a programmable, automatic instrument for carrying out the various stages of the method (Edman and Begg, 1967). The peptide or protein is spread by centrifugal force as a thin film on the inner wall of a spinning cup. The amount of substrate required for sequencing can be diminished by using a polymeric quaternary ammonium salt ('Polybrene') (5.21) which adheres strongly both to the substrate and to glass and effectively immobilises the former almost as in the solid-phase method (see below). The 2-anilinothiazolin-5-ones resulting initially are obtained separately by using a fraction collector and subsequently are isomerised to the 3-phenyl-2-thiohydantoins for identification. It is possible to accomplish 40–60 cycles with this apparatus. The amount of substrate required has been decreased further by using gaseous reagents at crucial points in the process. The sample solution is applied to a disc of glass filter paper coated with 'Polybrene'. As little as 10 pmol of substrate is enough to complete about twenty cycles, whereas the higher yields obtained with 10 nmol permit the identification of about ninety residues (Hunkapiller *et al.*, 1984). The development of the solid-phase method of peptide synthesis (Section 7.9) led to a search for a similar technique for the determination of amino-acid sequences. Separation of excess phenyl isothiocyanate and 2-anilinothiazolin-5-ones from the immobilised peptide is very simple. Most of the problems have centred around (a) finding the best insoluble support and (b) developing suitable methods for attaching the peptide to be sequenced. Aminopolystyrene was the first matrix to be tried, but it does not swell in aqueous media so that large peptides cannot obtain access. Consequently, coupling of the peptide had to be carried out in organic solvents in which it was usually poorly soluble and low yields resulted. An alternative type of support is based on polyacrylamides (Section 7.9) since these swell in water and peptides can be coupled to them in good yield. Finally, peptides can be immobilised covalently on controlled-pore glass. The latter is first treated with 3-aminopropyltriethoxysilane (5.22 →5.23) to provide an anchorage point (Scheme 5.4). The method of Hunkapiller *et al.* (1984) mentioned above can be regarded as a solid-phase method in which the use of 'Polybrene' renders unnecessary the covalent attachment of the peptide to the insoluble support.

Three main methods of coupling are used to attach peptides covalently to those

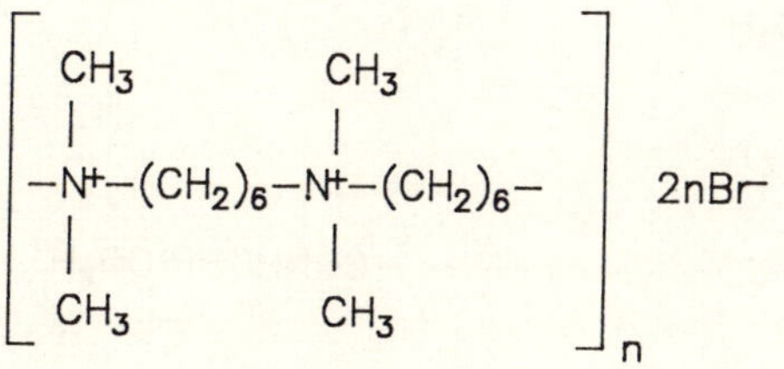

$$\left[\begin{array}{c} \overset{\overset{\displaystyle CH_3}{|}}{} \quad\quad \overset{\overset{\displaystyle CH_3}{|}}{} \\ -N^+-(CH_2)_6-N^+-(CH_2)_6- \\ \underset{\underset{\displaystyle CH_3}{|}}{} \quad\quad \underset{\underset{\displaystyle CH_3}{|}}{} \end{array}\right]_n \quad 2nBr^-$$

(5·21)

$$-O-\overset{\displaystyle /OH}{\underset{\displaystyle \backslash OH}{Si}}-OH \;+\; \overset{\displaystyle EtO}{\underset{\displaystyle EtO}{}}\overset{\displaystyle \backslash}{/}Si(CH_2)_3NH_2 \;\longrightarrow\; -O-\overset{\displaystyle /O\backslash}{\underset{\displaystyle \backslash O/}{Si}}-O-Si(CH_2)_3NH_2$$

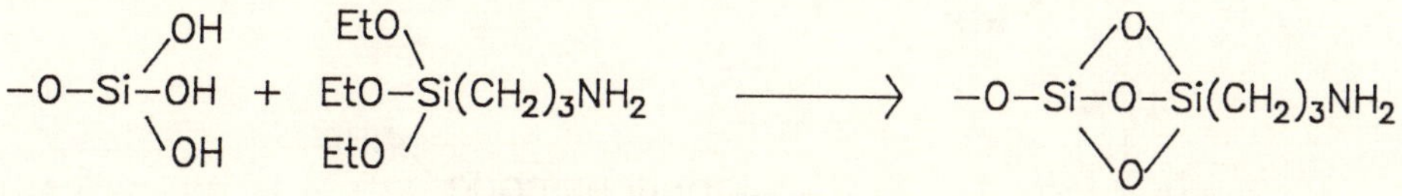

(5·22)   (5·23)

Scheme 5.4.

supports that contain free amino groups. In the first method, the peptide is treated with a water-soluble carbodiimide (Section 7.8) such as *N*-ethyl-*N'*-(3-dimethyl-amino)propylcarbodiimide in the absence of a nucleophile. The carboxy groups in the peptide initially give *O*-acylisoureas (Scheme 5.5) and those derivatives on the side-chains of aspartic and glutamic residues readily undergo intramolecular rearrangement to stable *N*-acylureas. The *O*-acylisourea at the *C*-terminus, however, preferentially cyclises to an oxazolin-5-one (5.24). This readily undergoes nucleophilic attack by the amino groups in the insoluble support, leading to ring-opening and the covalent coupling of the peptide derivative. It should be noted that Asp and Glu will appear as the derivatives (5.25; $n = 2,3$) in subsequent cycles of the Edman degradation. Since the peptide is attached to the support through its *C*-terminal residue, it should remain attached until the whole peptide has been sequenced, provided that the length of the peptide is within the limits of the method.

The second method also achieves attachment through the *C*-terminal residue. After cleavage of a polypeptide with cyanogen bromide (Section 5.8), all peptide fragments except that emanating from the *C*-terminus will end with a residue of homoserine (5.26) or its lactone (5.27). Complete lactonisation can be achieved by treatment with trifluoroacetic acid and the peptides then react directly with the amino groups on the support to give (5.28). Similarly, fragments from specific cleavage at Tyr or Trp residues contain spirolactones (Section 5.8), which can be coupled directly to a support bearing a free amino group.

In the third method, the amino groups on the support are brought into reaction with an excess of 4,4'-phenylene diisothiocyanate and the residual free isothiocyanate groups are available to couple with the α- and ε-amino groups on the peptide (Scheme 5.6). Treatment with trifluoroacetic acid cleaves the first peptide

101

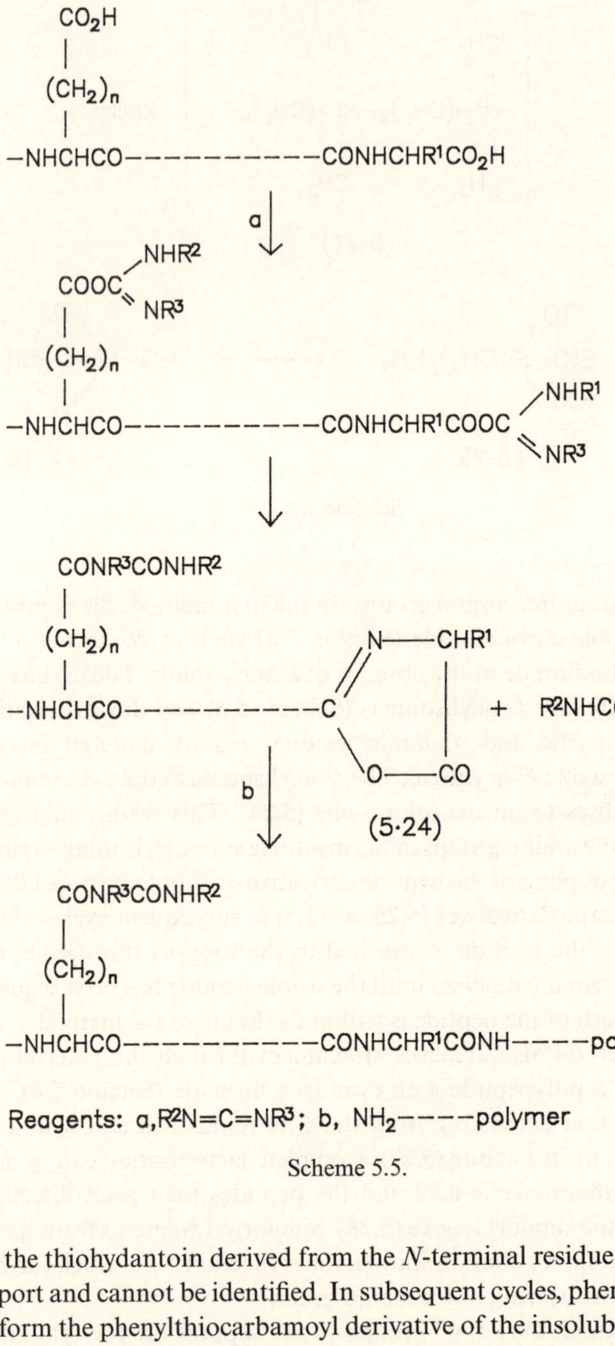

$$CO_2H$$
$$|$$
$$(CH_2)_n$$
$$|$$
$$-NHCHCO-------CONHCHR^1CO_2H$$

a ↓

$$COOC \overset{NHR^2}{\underset{NR^3}{\diagup}}$$
$$|$$
$$(CH_2)_n$$
$$|$$
$$-NHCHCO------CONHCHR^1COOC \overset{NHR^1}{\underset{NR^3}{\diagup}}$$

↓

$$CONR^3CONHR^2$$
$$|$$
$$(CH_2)_n$$
$$|$$
$$-NHCHCO-------C \begin{array}{c} N----CHR^1 \\ | \quad\quad | \\ O----CO \end{array} + R^2NHCONHR^3$$

b ↓ (5·24)

$$CONR^3CONHR^2$$
$$|$$
$$(CH_2)_n$$
$$|$$
$$-NHCHCO-------CONHCHR^1CONH----polymer$$

Reagents: a, $R^2N=C=NR^3$; b, $NH_2----polymer$

Scheme 5.5.

bond, but the thiohydantoin derived from the *N*-terminal residue remains attached to the support and cannot be identified. In subsequent cycles, phenyl isothiocyanate is used to form the phenylthiocarbamoyl derivative of the insolubilised peptide and the 3-phenyl-2-thiohydantoins formed by treatment with acid are collected and identified. Clearly no thiohydantoin will be liberated when the stepwise degradation reaches a Lys residue through which the peptide is attached to the support. If this is

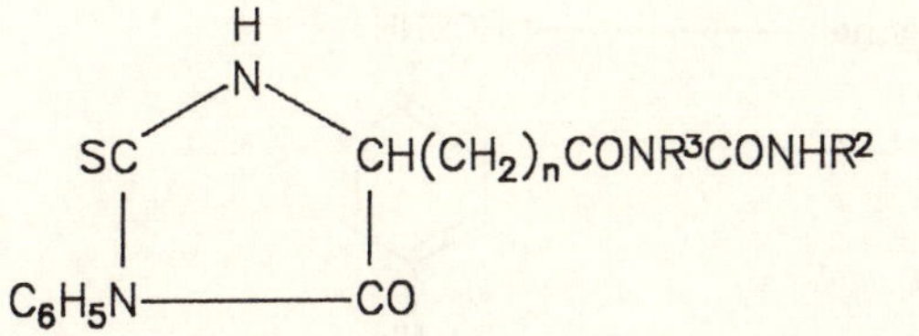

SC NH CH(CH₂)ₙCONR³CONHR² / C₆H₅N──CO

$$\text{(5·25)}$$

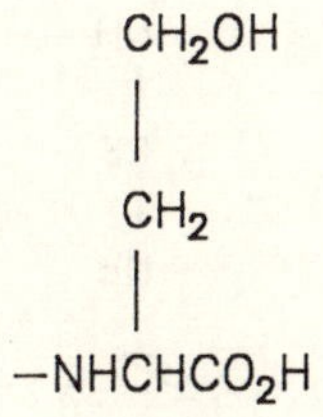

$$\begin{array}{c} CH_2OH \\ | \\ CH_2 \\ | \\ -NHCHCO_2H \end{array}$$

$$\text{(5·26)}$$

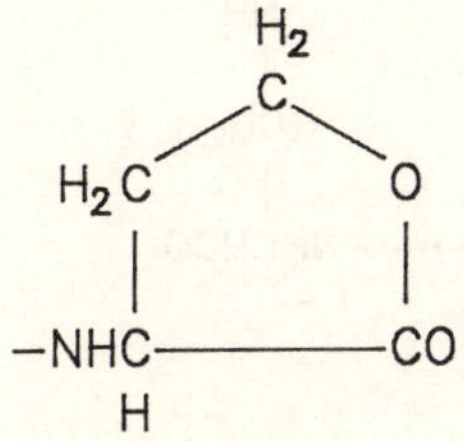

$$\text{(5·27)}$$

$$\begin{array}{c} CH_2OH \\ | \\ CH_2 \\ | \\ -NHCHCONH\cdots\text{polymer} \end{array}$$

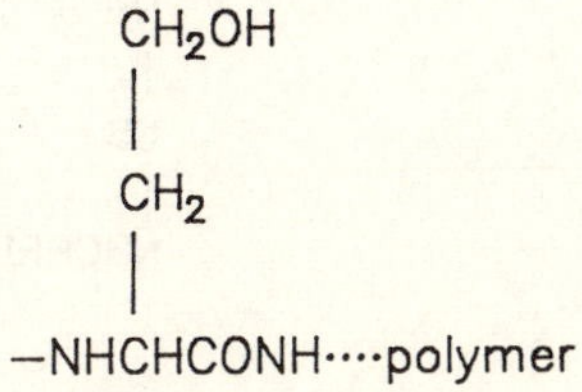

$$\text{(5·28)}$$

the only point of attachment, the remainder of the peptide falls off the support at this stage. Consequently, the maximum amount of information can be obtained if the C-terminal residue in the peptide is Lys. Fortunately, peptides with C-terminal lysine can be obtained by hydrolysing the original polypeptide with trypsin (Section 5.9). Trypsin also liberates peptides with C-terminal arginine, which has a guanidino group in the side-chain. Treatment of such peptides with 50% aqueous hydrazine at 75 °C for 15 min converts the C-terminal arginine into ornithine, albeit in rather low yield. Since ornithine contains a δ-amino group, these peptides can be coupled to the support just like lysine-containing peptides.

After polypeptides have been separated by polyacrylamide electrophoresis, they can be electrophoretically transferred to a glass-fibre filter paper that has been treated with 3-aminopropyltriethoxysilane and covalently bonded using 4,4′-phenylene diisothiocyanate. The glass-fibre disc loaded with as little as 10 μg of polypeptide can be placed directly into a gas-phase sequenator (Aebersold et al., 1988).

Unfortunately, the 3-phenyl-2-thiohydantoins formed in the Edman stepwise degradation suffer racemisation (Davies and Mohammed, 1984) so that the method cannot be used to determine the configuration of amino acids in a peptide. This is not usually a serious limitation, but enantiomerisation is a perpetual hazard in peptide synthesis (Chapter 7). It is therefore desirable to determine if enantiomerisation has occurred at any residue. Such information could be important, for example, in dating bone proteins obtained in archaeological excavations. Examination of the chiral purity of the amino acids in a total acid hydrolysate is not satisfactory.

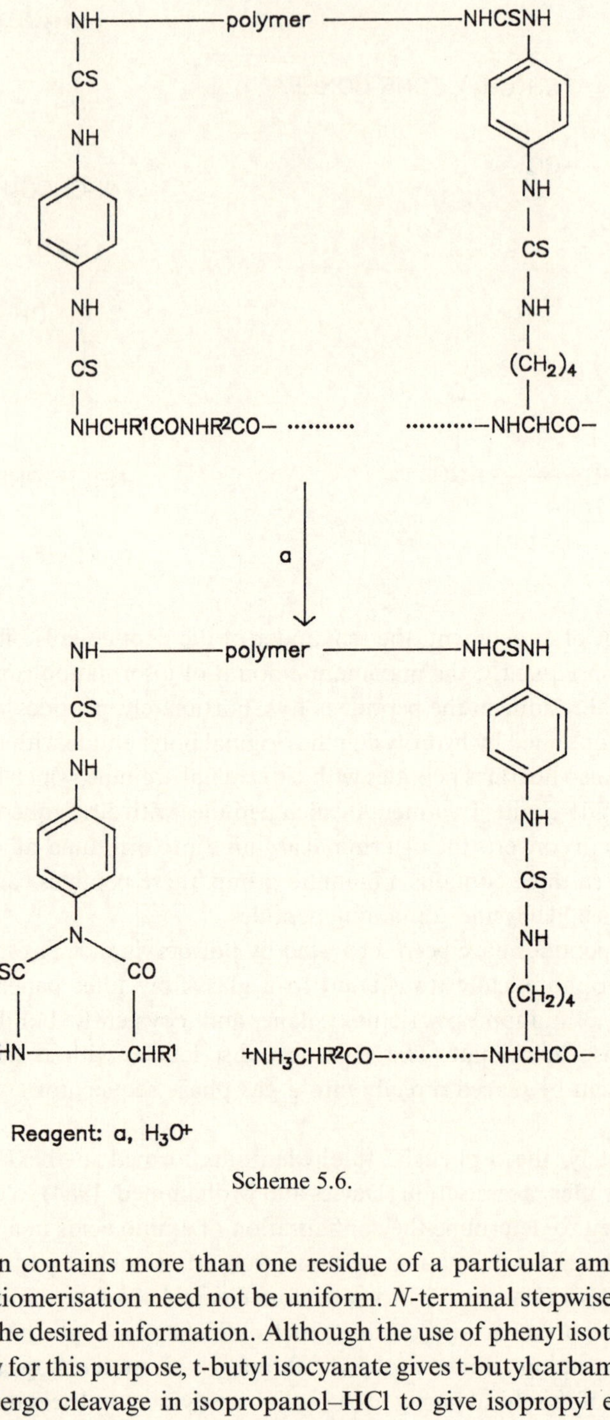

NH——————polymer——————NHCSNH
|
CS
|
NH
[benzene ring]
NH
|
CS
|
NHCHR¹CONHR²CO— ·········· ··········—NHCHCO—
$(CH_2)_4$
|
NH
|
CS
|
NH
[benzene ring]
NHCSNH

↓ a

NH——————polymer——————NHCSNH
|
CS
|
NH
[benzene ring]
N
SC   CO
| |
HN————CHR¹ $^+NH_3CHR^2CO$——············—NHCHCO—
$(CH_2)_4$
|
NH
|
CS
|
NH
[benzene ring]
NHCSNH

Reagent: a, H_3O^+

Scheme 5.6.

If the protein contains more than one residue of a particular amino acid, the degree of enantiomerisation need not be uniform. *N*-terminal stepwise degradation could provide the desired information. Although the use of phenyl isothiocyanate is not satisfactory for this purpose, t-butyl isocyanate gives t-butylcarbamoyl peptides, which can undergo cleavage in isopropanol–HCl to give isopropyl esters of *N*-t-butylcarbamoylamino acids. These resist racemisation and can be identified by enantioselective gas chromatography in glass capillaries coated with a silicone to which L-valine-(S)-α-phenylethylamide is covalently attached (Bolte *et al.*, 1987). An excellent practical protocol for automated solid-phase sequencing with home-made hardware was given by Findlay *et al.* (1989). For those who have no special hardware, including HPLC equipment, the manual Edman method can be employed and each *N*-terminal residue can be identified by withdrawing a small sample and subjecting it to dansylation, hydrolysis and TLC. The majority of the sample (about 95% at each stage) is degraded by the Edman procedure ready for identification of the next *N*-terminal residue (Yarwood, 1989). Perhaps the ultimate development of the Edman procedure, especially in terms of sensitivity, involves iso-lation of the *N*-phenylthiocarbamoyl amino acids followed by their conversion into the 2-anilinothiazolin-5-ones (**5.18**) and then ring-opening of these with 4-aminofluorescein to give *N*-phenylthiocarbamoyl derivatives of amino-acylaminofluorescein (Farnsworth and Steinberg, 1993a, b). This methodology permits automated sequencing through at least thirty cycles with no more than 1 pmol of protein.

5.5 Enzymic methods for determining *N*-terminal sequences

Although the Edman method is by far the most common method of sequencing pep-tides from the *N*-terminus, some enzymic methods are used occasionally. Several aminopeptidases are available commercially, which differ in their specificities. One aminopeptidase from porcine kidney preferentially releases amino acids such as leucine with hydrophobic side-chains. This enzyme does not release *N*-terminal Arg or Lys or any amino acid that is followed by Pro. Another enzyme, aminopeptidase M, which is obtained from the microsomal fraction of porcine kidney cells, is less specific and perhaps more useful. It is advisable to examine aliquots of the hydro-lysate at intervals by chromatography to determine the order in which amino acids are being released.

Another type of enzyme, termed a dipeptidyl aminopeptidase, releases dipeptides rather than amino acids from the *N*-terminus. Cathepsin C is one such enzyme and it will remove dipeptides consecutively from the *N*-terminus of a peptide until either Lys or Arg becomes the *N*-terminal amino acid or until Pro is in position 2 or 3 in the chain. Thus two dipeptides, Asp—Arg and Val—Tyr are cleaved from angiotensin II:

Asp—Arg—Val—Tyr—Ile—His—Pro—Phe,

whereas bradykinin,

Arg—Pro—Pro—Gly—Phe—Ser—Pro—Phe—Arg,

105

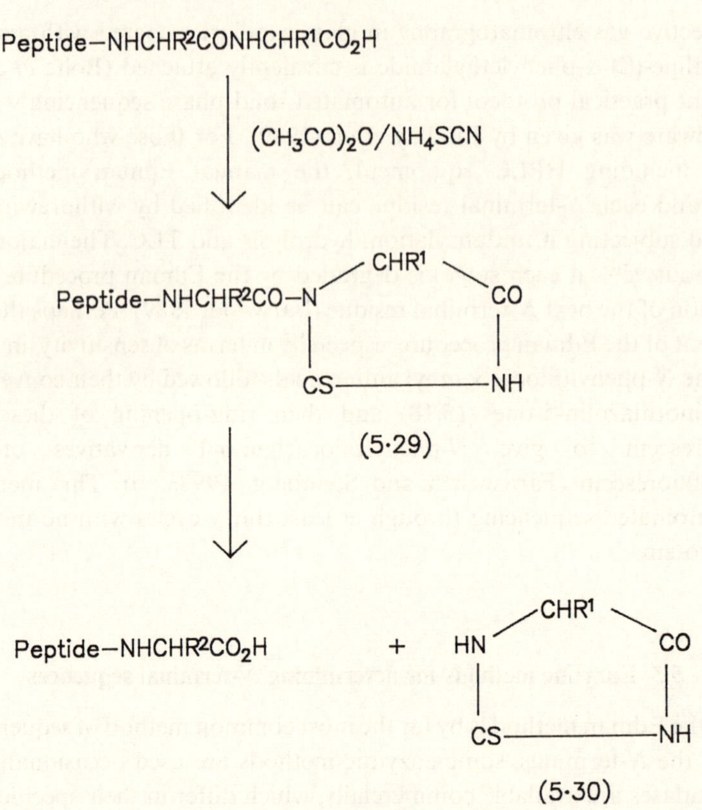

Peptide—NHCHR²CONHCHR¹CO₂H

$(CH_3CO)_2O/NH_4SCN$

Peptide—NHCHR²CO—N—CHR¹—CO / CS————NH

(5·29)

Peptide—NHCHR²CO₂H   +   HN—CHR¹—CO / CS————NH

(5·30)

Scheme 5.7.

is resistant. The dipeptides liberated by cathepsin C can be separated by ion-exchange chromatography or HPLC and identified by *N*-terminal and total amino-acid analysis. An alternative set of dipeptides can be isolated if the *N*-terminal residue is removed by one cycle of the Edman procedure before exposure to cathepsin C. This procedure can be useful if an *N*-terminal Arg or Lys is present or generated at an early stage with cathepsin C.

5.6 Identification of *C*-terminal sequences

It is many years since Schlack and Kumpf showed that a simple *N*-acyl peptide treated with ammonium thiocyanate and acetic anhydride (Scheme 5.7) underwent cyclisation at the *C*-terminus to yield 1-acyl-2-thiohydantoins (**5.29**). Mild alkaline hydrolysis then yielded the 2-thiohydantoin (**5.30**) corresponding to the *C*-terminal terminal residue and an *N*-acylpeptide containing one amino acid fewer. This reaction sequence should lead to a cyclic procedure at the *C*-terminus analogous to the Edman procedure at the *N*-terminus. Despite several attempts to avoid side reactions and improve yields, however, this method has not been developed beyond the stage of test sequencing of peptides of known structure.

5.7 Enzymic determination of *C*-terminal sequences

A limited amount of information can be obtained by the use of proteolytic enzymes that detach either amino acids or dipeptides sequentially from the *C*-terminus. They are thus complementary to the aminopeptidases and dipeptidyl aminopeptidases. Two pancreatic enzymes, carboxypeptidases A and B, differ in specificity. The former preferentially liberates *C*-terminal amino acids with aromatic side chains, somewhat less readily amino acids with alkyl side chains and, more slowly still, other amino acids, but not Pro, Arg, Lys and His. In contrast, carboxypeptidase B releases only *C*-terminal Arg, Lys and His. Carboxypeptidase Y is much less specific and is capable of removing all amino acids, although Gly and Pro are liberated only slowly. As with aminopeptidases, it is advisable to analyse the hydrolysate at intervals in order to determine the *C*-terminal sequence of amino acids. An interesting recent development (Carles *et al.*, 1988) uses carboxypeptidase to effect transpeptidation between the protein being sequenced and a tritiated amino acid. The labelled protein is then degraded by various specific methods and then the labelled fragments are isolated by gel electrophoresis and subjected to Edman degradation.

Dipeptidyl carboxypeptidases remove the *C*-terminal dipeptide intact and therefore are analogous to the dipeptidyl aminopeptidases such as cathepsin C. One such enzyme, angiotensin-converting enzyme, is important biologically for converting angiotensin I into the hypertensive angiotensin II (see Section 9.3). This enzyme does not hydrolyse bonds of the type X—Pro but will hydrolyse Pro—X bonds. The use of dipeptidyl carboxypeptidases for sequence determination would probably increase if pure enzymes were readily available commercially.

5.8 Selective chemical methods for cleaving peptide bonds

In order to obtain fragments from a protein of a suitable size for sequencing by the Edman method, it is desirable to effect specific cleavage adjacent to the rarer amino acids. Numerous chemical methods of specifically cleaving proteins have been devised, but only a few are in common use.

Cleavage of methionyl bonds occurs when proteins are allowed to react with cyanogen bromide (Gross and Witkop, 1961) (Scheme 5.8). The methionyl residue is converted into homoserine (**5.31**) and its lactone (**5.32**) and these form the *C*-terminus of all fragments except that which originates from the *C*-terminal segment of the protein. It should be noted that methionine is rather easily oxidised to the sulphoxide and this resists cleavage by CNBr. As mentioned above, the *C*-terminal homoserine lactone, which results from successful cleavage, provides a useful point of anchorage to an insoluble support for the solid-phase method of sequencing.

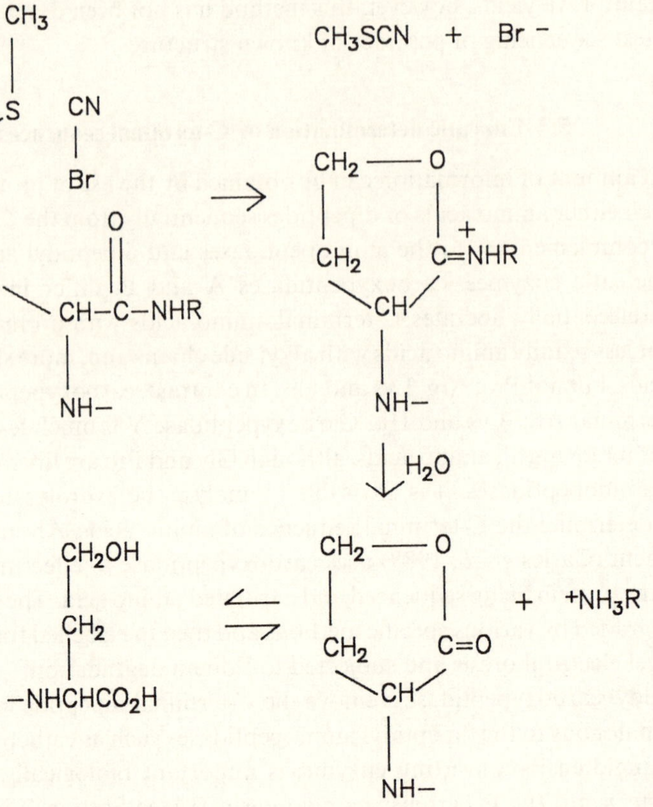

$$CH_3SCN \ + \ Br^-$$

CH$_3$
|
S   CN
| |
CH$_2$ Br
|
CH$_2$ O
| ‖
CH–C–NHR

NH–

→

CH$_2$——O
| |
CH$_2$ C=NHR$^+$
\   CH   /

NH–

↓ H$_2$O

CH$_2$OH
|
CH$_2$
|
–NHCHCO$_2$H

⇌

CH$_2$——O
| |
CH$_2$ C=O
\   CH   /

NH–

+ $^+$NH$_3$R

(5·31)   (5·32)

Scheme 5.8.

Tryptophan, like methionine, is a relatively rare amino-acid residue in proteins so that cleavage of tryptophyl bonds provides quite large fragments. The earliest method of cleavage used N-bromosuccinimide (Patchornik *et al.*, 1958) and this also caused cleavage of tyrosyl bonds. Although the latter is not particularly desirable since tyrosine is not a rare amino acid, the chemistry involved in the cleavage of this type of bond is slightly easier to understand and it will be described first. The aromatic ring is first brominated (Scheme 5.9) and the product (**5.33**) then undergoes a concerted nucleophilic attack by the carbonyl oxygen atom and displacement of a bromide ion to give the spirodienone (**5.34**). Hydrolysis of the $>C=N^+HR'$ bond effects the cleavage and the original Tyr residue forms a lactone (**5.35**) at the C-terminus of every fragment apart from the original C-terminal sequence. The final hydrolytic step resembles that in the method for cleaving proteins at methionyl bonds.

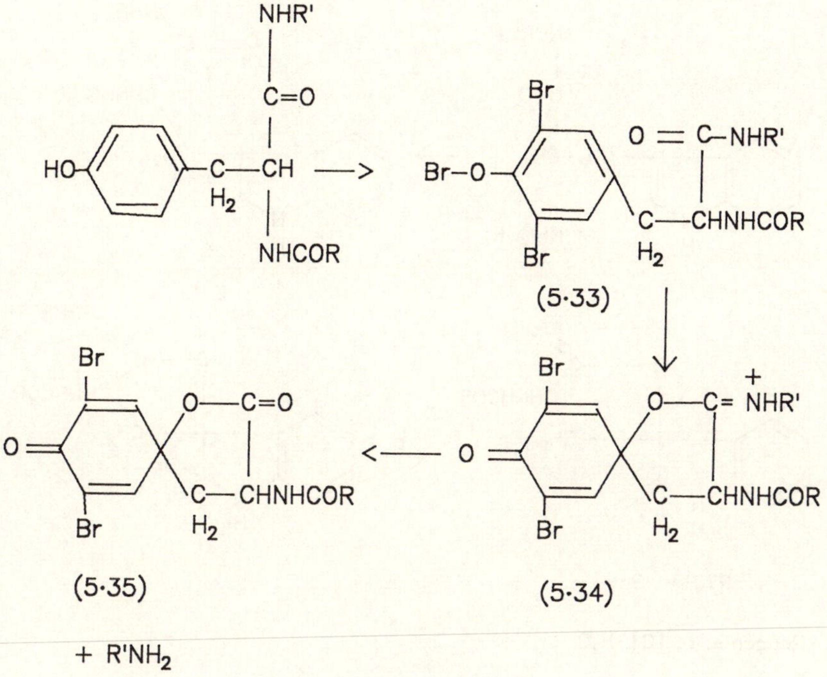

Scheme 5.9.

The sequence of events in the cleavage of tryptophyl bonds is somewhat similar (Scheme 5.10) with the exception that an oxidation step is involved. Two methods have been developed that are more selective towards tryptophyl bonds. These use either 2-(2-nitrophenylsulphenyl)-3-methyl-3-bromoindolenine (Omenn *et al.*, 1970; Fontana *et al.*, 1980) (**5.36**) (generated *in situ* from 2-(nitrophenylsulphenyl)-3-methyl-indole and *N*-bromosuccinimide) or 2-iodosobenzoic acid (Mahoney and Hermodson, 1979; Mahoney *et al.*, 1981) (**5.37**). Both reagents probably react by a mechanism similar to that with *N*-bromosuccinimide. 2-Iodosobenzoic acid requires the addition of a halide ion and it is convenient to use guanidinium chloride since this denatures the protein and affords higher yields of cleavage products. Oxidative chlorination of the indole ring of Trp and of the phenol group of Tyr takes place. The cleavage can be selectively limited to Trp residues if desired by adding an excess of 4-cresol as a competitive scavenger to protect Tyr groups.

Several other methods for the selective chemical cleavage of peptide bonds adjacent to particular amino acids have been described, but none has been used widely.

5.9 Selective enzymic methods for cleaving peptide bonds

It was mentioned above (Section 5.4) that trypsin cleaves lysyl and arginyl bonds so that, when a protein is exhaustively degraded by this enzyme, only one fragment can

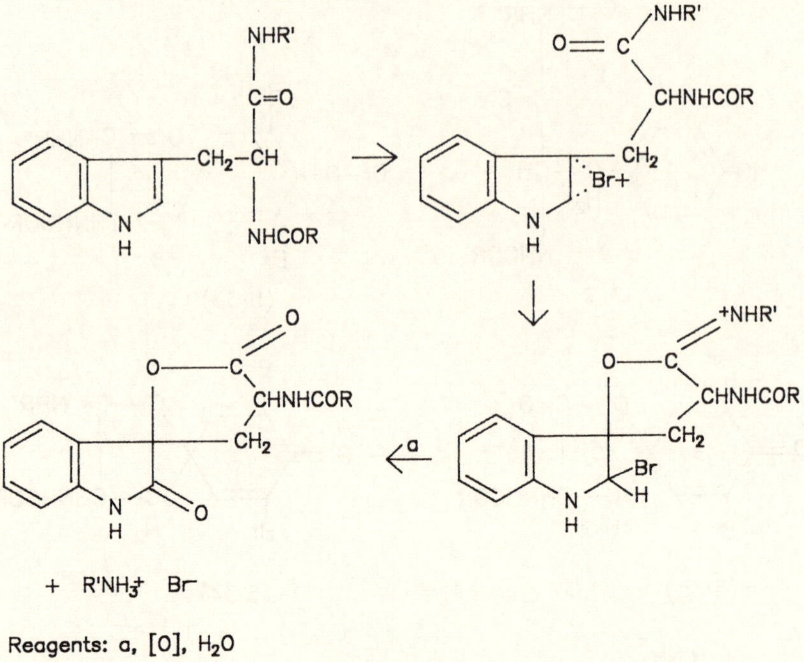

Reagents: a, [0], H₂0

Scheme 5.10.

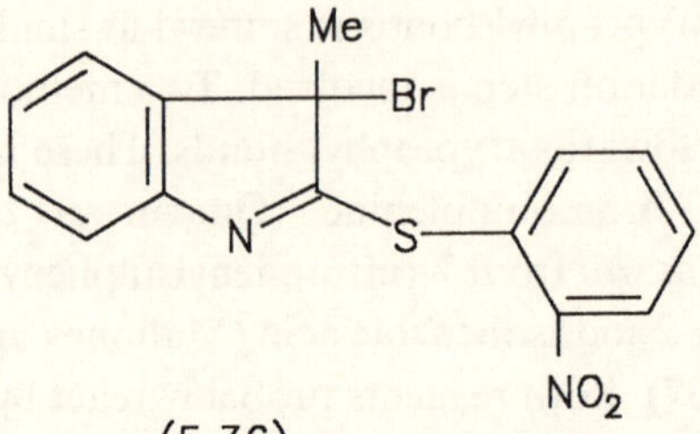

(5·36)

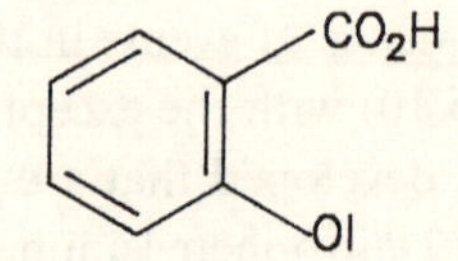

(5·37)

have a *C*-terminal residue differing from Lys or Arg. That unique fragment contains the *C*-terminal sequence of the protein. It is possible, of course, that the *C*-terminal residue of the original protein is Lys or Arg and then it is impossible to identify the *C*-terminal fragment from tryptic hydrolysis. In this case, the *C*-terminal residue can be removed from the original protein using carboxypeptidase B before degrading the protein with trypsin. This procedure will ensure that the *C*-terminal fragment can be identified by the absence of Lys or Arg. Alternatively, instead of removing the *C*-terminal residue, it can be exchanged for the radioactive residue by incubation of the protein with the appropriate ¹⁴C-labelled amino acid (Lys or Arg) in the presence of carboxypeptidase B (Charles *et al.*, 1988). When the protein is hydrolysed by trypsin, only the *C*-terminal peptide will be radioactive.

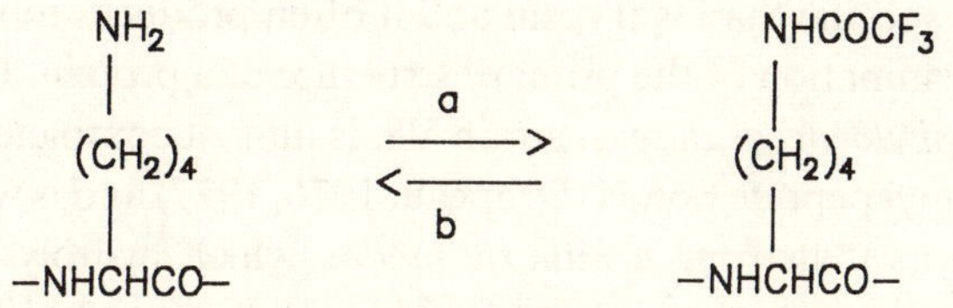

NH$_2$ NHCOCF$_3$
| |
(CH$_2$)$_4$ $\xrightarrow{a}$ (CH$_2$)$_4$
| $\xleftarrow{b}$ |
—NHCHCO—   —NHCHCO—

Reagents: a, CF$_3$CO$_2$C$_2$H$_5$; b, morpholine

Scheme 5.11.

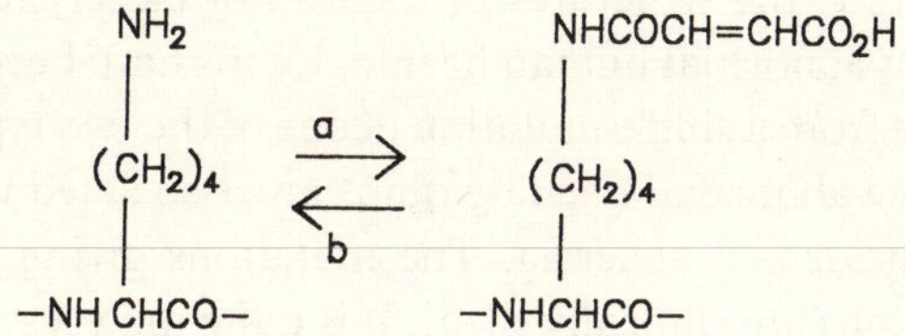

NH$_2$ NHCOCH=CHCO$_2$H
| |
(CH$_2$)$_4$ $\xrightarrow{a}$ (CH$_2$)$_4$
| $\xleftarrow{b}$ |
—NH CHCO—   —NHCHCO—

Reagents: a, maleic anhydride; b, H$_3^+$O (pH 2)

Scheme 5.12.

When determining the primary structure of a basic protein that affords an inconveniently large number of tryptic fragments, it is possible to restrict the cleavage to arginyl bonds by first protecting the ε-amino groups of the Lys residues. Since trypsin recognises its specific substrates by the presence of a positively charged side-chain, any acylating reagent will suffice but two are particularly useful because they can be removed subsequently to expose the lysine side-chain again for a second hydrolysis with trypsin. ε-Amino groups can be trifluoroacetylated with ethyl trifluoroacetate and deprotected with a base such as morpholine (Scheme 5.11). Alternatively, the Lys side-chains can be maleylated using maleic anhydride. In this case, the temporary protecting groups can be removed by very mild acid treatment (Scheme 5.12).

The use of thrombin, an enzyme that serves several roles in the blood-coagulating process, is a useful adjunct to tryptic hydrolysis. Its action is more specific and it cleaves only a limited number of arginyl bonds as a rule. Some arginyl bonds are only slowly hydrolysed by thrombin so that enzymic digests of protein can be quite complex in composition because degradation of substrate is incomplete.

Chymotrypsin gives an alternative set of peptides because cleavage occurs at those peptide bonds which contain the carbonyl group of aromatic amino acids (Phe, Trp and Tyr) or hydrophobic aliphatic amino acids (Leu, Met and Ile). Clearly, chymotrypsin is less specific than is trypsin and it often produces too many small peptides for the determination of the primary structure of a protein. In contrast, a proteinase from *Staphylococcus aureus*, strain V8, is almost completely specific for the cleavage of glutamyl peptide bonds (Drapeau, 1976, 1977) and is widely used. Other proteinases such as that from *Armillaria mellia*, which hydrolyses peptide bonds containing the α-nitrogen atom of Lys residues (Lewis *et al.*, 1978), and another from the yeast *Candida tropicalis*, which is specific for the hydrolysis of valyl bonds (Abassi *et al.*, 1986), would be extremely useful if they were more easily available.

Other less specific proteinases such as papain, subtilisin and pepsin are mainly of value for isolating small peptides containing disulphide bonds, phosphoserine or amino acids bearing carbohydrate attachments in their side-chain.

Selective enzymic hydrolysis of peptide bonds is particularly time-saving when it is required to determine the structures of a group of closely related proteins. For example, hundreds of abnormal human haemoglobins have been discovered. These proteins usually arise from a single mutation in one of the two types of chain present in the molecule. Some abnormal haemoglobins are associated with serious clinical conditons such as sickle-cell anaemia. The mutations giving rise to some other abnormal haemoglobins are clinically silent. It is quite unnecessary to follow all the procedures described above in order to determine the molecular site of a single mutation. If a protein of known sequence and a closely related protein are subjected to hydrolysis with a fairly specific proteinase such as trypsin, most of the peptides resulting from enzymic digestion will be revealed to be identical by HPLC, two-dimensional TLC and electrophoresis. The pattern of peaks or spots obtained is often referred to as a tryptic fingerprint. It is only necessary to determine the structure of those peptides which are different in the hydrolysates of the two proteins. Thus the *N*-terminal octapeptide obtained from the tryptic hydrolysis of the β-chain of normal adult haemoglobin (HbA) and the corresponding peptide from the haemoglobin which is present in patients with sickle-cell anaemia (HbS) differ at residue 6:

> HbA β-chain: Val—His—Leu—Thr—Pro—*Glu*—Glu—Lys
>
> HbS β-chain: Val—His—Leu—Thr—Pro—*Val*—Glu—Lys.

The remainder of the β-chain and all of the α-chain are identical in HbA and HbS. Several experimental protocols for mapping proteins have been described (Carrey, 1989).

5.10 Determination of the positions of disulphide bonds

Insulin contains three disulphide bonds, namely two interchain disulphide bonds and one intrachain structure in the A chain. These disulphide bonds tend to undergo slow exchange reactions when exposed to acid:

$$RSSR + H^+ \rightleftarrows RS^+ + RSH$$
$$RS^+ + R'SSR' \rightleftarrows RSSR' + R'S^+$$
$$R'S^+ + RSSR \rightleftarrows RSSR' + RS^+.$$

In addition, if a protein contains a thiol group and a disulphide bond, slow exchange can occur under basic conditions:

$$RSH \rightleftarrows RS^- + H^+$$
$$RS^- + R'SSR' \rightleftarrows RSSR' + R'S^-$$
$$R'S^- + RSSR' \rightleftarrows RSSR + R'S^-.$$

The classical approach to the location of disulphide bonds involves hydrolysis of the protein under conditions such that the risk of disulphide-bond exchange is minimised. A proteinase of low specificity such as pepsin or thermolysin is likely to yield products of a suitable size. Papain may be unsuitable since it contains a free thiol group that could catalyse exchange reactions. The proteolytic fragments can be separated by paper electrophoresis, exposed to performic acid vapour in order to oxidise cystine residues to cysteic acid and finally subjected to electrophoresis under the same conditions as before but with the direction of current flow rotated through 90°. All peptides not containing cystine will be found on a diagonal since they migrate for equal distances at right angles. Peptides that did contain cystine will have been made more acidic as a result of oxidation to cysteic acid. Consequently, the fragments will be displaced from the diagonal in a direction towards the anode for the second separation. Interchain disulphides should give two new spots but complete resolution may not always be achieved. Intrachain disulphides will give only one new product (Brown and Hartley, 1966; Aitken et al., 1988; Creighton, 1989), except when proteolyic cleavage has occurred within the disulphide loop during the attempt to locate it. The great resolving power of HPLC has provided an alternative method (Lu et al., 1987). The protein is subjected to hydrolysis by a suitable proteinase, the peptides are separated by reversed-phase HPLC and the amino-acid composition is determined (Chapter 3). The peptide fragments that contain cystine are then subjected to Edman degradation. Although two thiohydantoins are usually formed at each step because there are two chains, location of the sequences within the predetermined primary structure of the protein is straightforward. This method can be used on a few picomoles of the protein.

Finally, disulphide bonds can be located by hydrolysing a protein to a mixture of peptides using either a proteinase or a specific chemical method of cleavage and the mixture can be analysed directly by fast-atom bombardment mass spectrometry (Chapter 3) and again after reduction of disulphide bonds (Yazdanparast et al., 1987). By identifying those peaks which disappear as a result of reduction and new peaks with appropriate masses that have taken their place, it is simple to assign disulphide bonds to the relevant amino-acid sequences.

5.11 Location of post-translational modifications and prosthetic groups

Some examples of crosslinks that arise by post-translational modification of proteins have been given above. Other possible changes to amino-acid residues include acetylation of *N*-terminal α-amino groups, formation of *C*-terminal amide groups, phosphorylation or sulphation of hydroxy groups, methylation of ε-amino groups of lysine or γ-carboxylation of glutamic acid. The location of modified amino acids involves degrading the protein to small fragments like when locating disulphide bonds and then identifying the sequence containing the non-coded amino acid. The conditions for effecting the degradation must not further modify the structure. For example, γ-carboxy-glutamic acid is decarboxylated very easily under acidic conditons to give glutamic acid. Identification of this amino acid in prothrombin was delayed because the amino-acid composition determined after acid hydrolysis was normal.

Many proteins and especially enzymes contain a moiety that is not peptidic in nature. Identification and location of such a structure may be facilitated by characteristic light-absorption properties. Frequently, such a moiety is essential to the biological activity of the protein and it is then referred to as a prosthetic group. For example, some enzymes that carboxylate substrates such as pyruvate, acetyl-CoA and propionyl-CoA have a molecule of biotin (**5.38**) covalently linked as an amide to the enzyme through the ε-amino group of a lysine residue. There are instances in which a non-peptidic moiety is bound tightly but non-covalently to a protein. A good example is the haem component of haemoglobin. The haem is retained during dialysis but dissociates on treatment with acid. Degradative methods cannot be used to locate non-covalently bound molecules and resort has to be made to techniques such as X-ray- and neutron-diffraction methods.

A particularly interesting example of a prosthetic group is found with transaminases. These enzymes catalyse the transfer of an amino group to an α-keto acid (see Chapter 8). It is sufficient to state here that the coenzyme for the reaction is alternately pyridoxal phosphate (**5.39**) and pyridoxamine (**5.40**). The former is covalently bound to the enzyme as an aldimine involving the ε-amino group of a lysine residue whereas the latter is only bound non-covalently to the enzyme. The existence of the aldimine can be demonstrated by reducing it to the corresponding secondary amine with sodium borohydride. This treatment prevents the subsequent enzyme-catalysed steps and inactivates the enzyme. If the reduction is effected with NaB^3H_4, the radioactive marker facilitates the identification of relevant peptides in degradative studies on the inactivated enzyme.

Again in the first step of the reaction between 1,3-dihydroxyacetone monophosphate (**5.41**) and glyceraldehyde-3-phosphate catalysed by aldolase to form fructose-1,6-diphosphate or the reverse reaction, a ketimine (**5.42**) is formed between the substrate and the ε-amino group of a Lys residue in the enzyme. The formation of this intermediate (**5.42**) can be demonstrated similarly by trapping it as a secondary amine using $NaBH_4$ to reduce the ketimine. The glyceraldehyde-3-

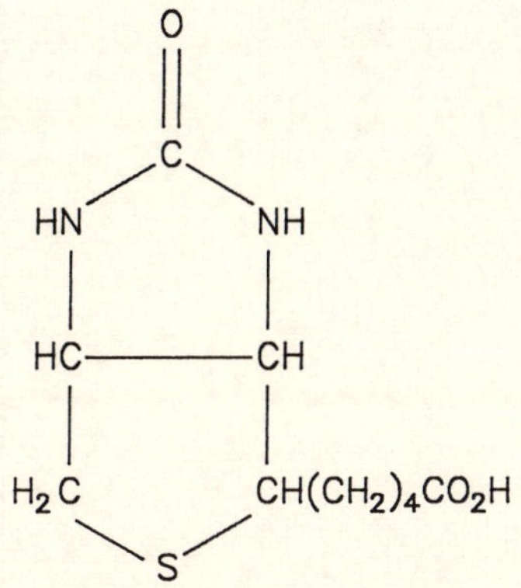

(5.38)

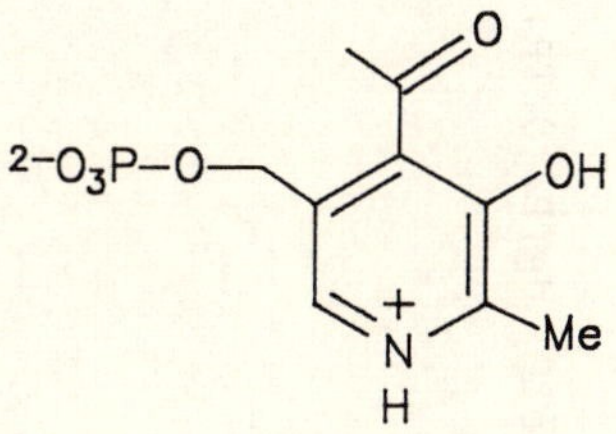

(5.39)

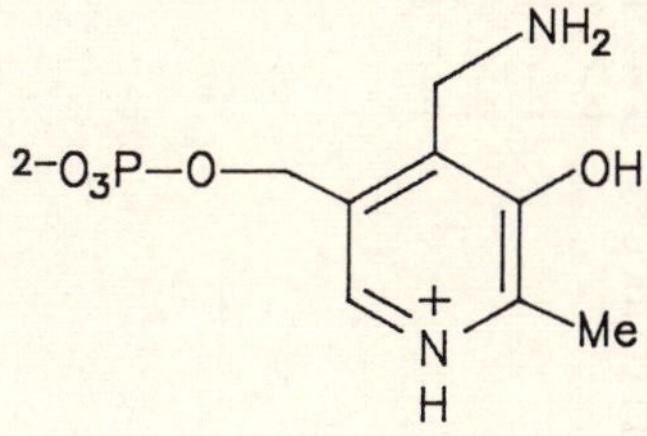

(5.40)

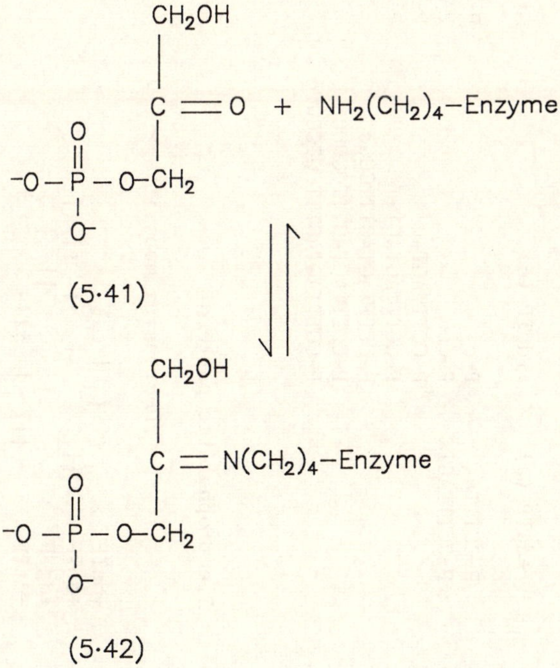

(5.41)

(5.42)

115

Single-stranded DNA template

```
3' ...................GGAAATCATGAAAGCCGGCAG....5'
5' -------------3' -OH
   Primer (P)
```

+dATP, dCTP, dGTP, dTTP + Klenow fragment
+ONE of the 2' 3' -dideoxyribonucleoside-5' -triphosphates (daTP, dcTP, dgTP, dtTP)

+dATP (a)	+dcTP (c)	+dgTP (g)	+dtTP (t)
P-CCTTTa	P-c	P-CCTTTAg	P-CCt
P-CCTTTAGTa	P-Cc	P-CCTTTAGTACTTTCg	P-CCTt
	P-CCTTTAGTAc	P-CCTTTAGTACTTTCGg	P-CCTTt
	P-CCTTTAGTACTTTc	P-CCTTTAGTACTTTCGGCCg	P-CCTTTAGt
	P-CCTTTAGTACTTTCGGc		P-CCTTTAGTACt
	P-CCTTTAGTACTTTCGGCc		P-CCTTTAGTACTt
	P-CCTTTAGTACTTTCGGCCGTc		P-CCTTTAGTACTTt
			P-CCTTTAGTACTTTCGGCCGt

Electrophoretic pattern

Decreasing mobility ----->

```
+daTP
+dcTP
+dgTP
+dtTP
```

```
5' ...CCTTTAGTACTTTCGGCCGTC...3'   Sequence of DNA synthesized on primer
3' ...GGAAATCATGAAAGCCGGCAG...5'   Sequence of template, single-stranded DNA
   GlyAsnHisGluSerArgGln           Amino-acid sequence
```

Figure 5.1. The Sanger method for sequencing DNA. In the four columns are listed the oligonucleotides formed in the presence of radioactive dideoxynucleoside phosphates. The electrophoretic pattern shows the relative positions of the four types of oligonucleotides. Decreasing mobility is equated with increasing molecular size.

phosphate first binds non-covalently to the enzyme and then condenses with the carbanion which is formed from (5.42) by loss of a proton.

5.12 Determination of the sequence of DNA

Determination of the sequence of DNA might be thought to be a slow method to determine the primary structure of a protein because three nucleotides have to be identified to discover the corresponding amino acid. The fastest method (Sanger, 1981) for determining the structure of DNA can educe sequences of several hundred nucleotides per working day. Even when this figure is divided by three, the number of amino-acid residues identified in the resultant protein is much greater than can be achieved by any of the variants of the Edman method. The reader may wonder why the direct sequencing of peptides and proteins is still undertaken. Most of the reasons have already been given. One of the most definitive methods for confirming the structure of a synthetic peptide is to determine its amino-acid sequence by degradative methods. Secondly, the DNA structure contains no information about the nature and sites of post-translational modifications. The frequency of occurrence of disulphide-bond formation underlines the continuing importance of the methods described so far. It must also be mentioned that the DNA approach carries with it the need to identify the open reading frame in the DNA structure (Stormo, 1987). On the other hand, determination of the sequence of some DNA molecules has given the primary structure of proteins that had not been isolated up to that point.

Sanger's method for determining the nucleotide sequence of DNA depends on making partial copies of the DNA in the single-stranded form on a single-stranded primer (Sanger *et al.*, 1977; Smith, 1980). Copying of the single strand is effected by the Klenow fragment of the DNA polymerase I from *E. coli* using a mixture of the four 2'-deoxyribonucleoside-5'-triphosphates, one of which is heavily labelled with ^{32}P on the α-phosphate group. The Klenow fragment lacking the 5' to 3' exonuclease is used to prevent attack on the 5' end of the primer. In addition, one of the four possible 2',3'-dideoxyribonucleoside-5'-triphosphates is present in the digest. Copying terminates in a Monte Carlo fashion by the incorporation of the 2',3'-dideoxyribonucleoside-5'-triphosphate in place of the corresponding 2'-deoxyribonucleoside-5'-triphosphate. The enzyme is able to couple these quasi-substrates at the 3' terminus of the growing DNA, but the absence of a 3'-hydroxy group prevents the extension of the chain.

The requirement for a single-stranded form of DNA could have been a serious handicap, but fortunately it proved possible to use the single-stranded M13 bacteriophage as a vector. The DNA to be sequence is sub-cloned into the double-stranded replication form of the M13 bacteriophage from which the single-stranded form can easily be prepared. An added bonus is gained because the cloning procedure is also an effective purification process (Barnes *et al.*, 1983; Messing, 1983). A short piece of complementary oligonucleotide primer is chemically synthesised and segments of the complementary strand are built on to this using the DNA polymerase as described above (see also Figure 5.1).

The method also relies on the high-resolution electrophoresis on denaturing poly-acrylamide gels to resolve polynucleotides with one common end (the primer) but varying in length at the other end by one nucleotide residue. As a result of the Monte Carlo process for termination of DNA copying, gel electrophoresis produces a series of oligonucleotide ladders depending on the length of the copy and which of the 2'-3'-dideoxyribonucleoside-5'-triphosphates is present in the digest. The derivation of the sequence from the electrophoresis pattern is illustrated in Figure 5.1.

5.13 References

Abassi, A., Voelter, W. and Zaidi, Z. H. (1986) *Biol. Chem. Hoppe-Seyler*, **367**, 441.

Aebersold, R. H., Pipes, G. D., Nika, H., Hood, L. E. and Kent, S. B. H. (1988) *Biochemistry*, **27**, 6860.

Aitken, A., Geisow, M. J., Findlay, J. B. C., Holmes, C. and Yarwood, A. (1989) in *Protein Sequencing: A Practical Approach*, ed. J. B. C. Findlay and M. J. Geisow, IRL Press, Oxford, p. 43.

Barnes, W. M., Bevan, M. and Son, P. H. (1983) *Methods Enzymol.*, **101**, 98.

Bolte, T., Yu, D., Stuwe, H. T., König, W. A. (1987) *Angew. Chem., Int. Ed.*, **26**, 331.

Brown, J. R. and Hartley, B. S. (1966) *Biochem. J.*, **101**, 214.

Carles, C., Huet, J.-C. and Ribadeau-Dumas, B. (1988) *FEBS Lett.*, **229**, 265.

Carrey, E. (1989) in *Protein Structure: A Practical Approach*, ed. T. E. Creighton, IRL Press, Oxford, p. 117.

Creighton, T. E. (1989) in *Protein Structure: A Practical Approach*, ed. T. E. Creighton, IRL Press, Oxford, p. 155.

Davies, J. S. and Mohammed, A. K. (1984) *J. Chem. Soc., Perkin Trans.*, **2**, 1723.

Drapeau, G. R. (1976) *Methods Enzymol.*, **45**, 469.

Drapeau, G. R. (1977) *Methods Enzymol.*, **47**, 189.

Edman, P. (1949) *Arch. Biochem.*, **22**, 475.

Edman, P. (1950) *Acta Chem. Scand.*, **4**, 283.

Edman, P. and Begg, G. (1967) *Eur. J. Biochem.*, **1**, 80.

Elmore, D. T. (1961) *J. Chem. Soc.*, 3161.

Farnsworth, V. and Steinberg, K. (1993a) *Analyt. Biochem.*, **215**, 190.

Farnsworth, V. and Steinberg, K. (1993b) *Analyt. Biochem.*, **215**, 200.

Findlay, J. B. C., Pappin, D. J. C. and Keen, J. N. (1989) in *Protein Sequencing: A Practical Approach*, ed. J. B. C. Findlay and M. J. Geisow, IRL Press, Oxford, p. 69.

Fontana, A., Savige, W. E. and Zombonin, M. (1980) in *Methods in Peptide and Protein Sequence Analysis*, ed. C. Birr, Elsevier/North-Holland Biomedical Press, Amsterdam, p. 309.

Gross, E. and Witkop, B. (1961) *J. Amer. Chem. Soc.*, **83**, 1510.

Hunkapiller, M. W., Kent, S., Caruthers, M., Dreyer, W., Firca, J., Giffin, C., Horvath, S., Hunkapiller, T., Tempst, P. and Hood, L. (1984) *Nature*, **310**, 105.

Lewis, W. G., Basford, J. M. and Walton, P. L. (1978) *Biochim. Biophys. Acta*, **522**, 551.

Lu, H. S., Klein, M. L., Everett, R. R. and Lai, P.-H. (1987) in *Protein Structure and Function*, ed. J. I. L'Italien, Plenum Press, New York, p. 493.

Mahoney, W. C. and Hermodson, M. A. (1979) *Biochemistry*, **18**, 3810.

Mahoney, W. C., Smith, P. K. and Hermodson, M. A. (1981) *Biochemistry* **20**, 443.

Messing, J. (1983) *Methods Enzymol.*, **101**, 20.

Omenn, G. S., Fontana, A. and Anfinsen, C. B. (1970) *J. Biol. Chem.*, **245**, 1895.

Patchornik, A., Lawson, W. B. and Witkop, B. (1958) *J. Amer. Chem. Soc.*, **80**, 4747.

Sanger, F. (1981) *Science*, **214**, 1205.

Sanger, F., Nicklen, S. and Coulson, A. R. (1977) *Proc. Natl. Acad. Sci., U.S.A.*, **74**, 5463.

Smith, A. J. H. (1980) *Methods Enzymol.*, **65**, 560.

Stormo, G. D. (1987) in *Nucleic Acid and Protein Sequence Analysis*, ed. M. J. Bishop and C. J. Rawlings, IRL Press, Oxford, p. 231.

Yarwood, A. (1989) in *Protein Sequencing: A Practical Approach*, ed. J. B. C. Findlay and M. T. Geisow, IRL Press, Oxford, p. 119.

Yazdaparast, R., Andrews, P. C., Smith, D. L. and Dixon, J. E. (1987) *J. Biol. Chem.*, **262**, 2507.

6

Synthesis of amino acids

6.1 General

There is an abundant supply of L-enantiomers of most of the coded amino acids. These are made available through large-scale fermentative production in most cases, and also through processing of protein hydrolysates. The early sections of this chapter cover this aspect, However, laboratory synthesis methods are required for the provision of most of the other natural amino acids and for all other amino acids, so the main part of this chapter deals with established syntheses.

6.2 Commercial and research uses for amino acids

In addition to the provision of supplies of common amino acids, there are growing needs for routes to new amino acids, since pharmaceutically useful compounds of this class continue to be discovered, which must be free from toxic impurities and homochirally pure in this particular context. Important functions for close analogues of coded and other biologically significant amino acids include *enzyme inhibition* and retarding the growth of undesirable organisms (fungistatic, antibiotic and other physiological properties, possessed either by the free amino acids or by peptides containing them). Free amino acids that perform in this way are α-amino isobutyric acid (an example of an α-methylated analogue of a coded amino acid), which has been proposed for the control of domestic wood-rotting fungi), and α-methyl-Dopa (α-methyl-3′,4′-dihydroxy-L-phenylalanine), a well-known treatment for Parkinson's disease. Similar success for new therapeutic amino acids, based on their enzyme-inhibition properties, is indicated for amino acids with a minimal structural change such as the substitution of a side-chain hydrogen atom by a fluorine atom.

6.3 Biosynthesis: isolation of amino acids from natural sources

Many examples of the discovery and isolation of amino acids from natural sources date from the early 1900s, though some were characterised several years before that (Greenstein and Winitz, 1961). Further new examples continue to be discovered, either as constituents of proteins, revealing new post-translational processes for higher organisms (Table 1.3 in Chapter 1), or in the free or bound form (from fungal or bacterial sources or from marine organisms).

6.3.1 Isolation of amino acids from proteins

Hydrolysis of proteins and separation of the resulting mixture is an obvious, and traditional way (Greenstein and Winitz, 1961) of obtaining moderate quantities of the coded and post-translationally modified L-α-amino acids. However, because of the availability of viable methods of industrial synthesis, hydrolysis of proteins no longer offers a sensible approach owing to its tedious and expensive nature and the fact that some amino acids are destroyed in the process (see Chapter 3).

6.3.2 Biotechnological and industrial synthesis of coded amino acids

Knowledge gained of biosynthetic routes to L-α-amino acids and isolation of the enzymes mediating the steps in these routes has been exploited for the industrial-scale manufacture of most of the coded L-α-amino acids. In some cases, the enzymatic production of near-analogues of the coded L-α-amino acids can also be achieved (Goldberg and Williams, 1991; Rozzell and Wagner, 1992).

To illustrate the methods, a culture medium that contains indole, pyruvic acid, tyrosine phenollyase and an ammonium salt, as well as the usual buffers and salts, will accumulate L-tryptophan; or will produce an indole-substituted L-tryptophan if indole itself is replaced by a substituted indole. L-Dopa formed in a system employing tyrosinase from *Aspergillus terreus* provides a further example of this approach (Chattopadhyay and Das, 1990).

The crucial enzymes need not be isolated, since 'bio-reactors' containing micro-organisms that are fed with the appropriate starting materials are often more convenient. L-Threonine from *Brevibacterium flavum*, L-lysine from *Corynebacterium glutamicum* (Eggeling, 1994) and use of plant-cell suspension cultures illustrated by L-Dopa from *Mucuna pruriens* (Wichers *et al.*, 1985) are examples. However, bio-engineering of the whole organisms to be used in this way may need to be carefully optimised to achieve reasonable yields. The other main opportunity offered by biotechnological methods is the conversion of one amino acid into a less plentifully available amino acid, e.g. the conversion of L-tyrosine into L-Dopa using *Mucuna pruriens* (Wichers *et al.*, 1985).

For a limited range of amino acids, this approach is increasingly in competition

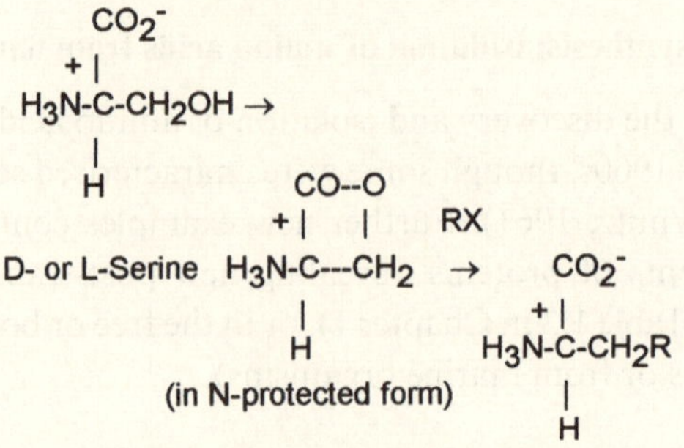

$$
\begin{array}{c}
CO_2^- \\
+| \\
H_3N\text{-}C\text{-}CH_2OH \rightarrow \\
| \\
H
\end{array}
\qquad
\begin{array}{c}
CO\text{-}O \\
+| \quad | \quad RX \\
H_3N\text{-}C\text{---}CH_2 \rightarrow \\
| \\
H
\end{array}
\qquad
\begin{array}{c}
CO_2^- \\
+| \\
H_3N\text{-}C\text{-}CH_2R \\
| \\
H
\end{array}
$$

D- or L-Serine   (in N-protected form)

Scheme 6.1.

with chemical synthesis, which can accomplish the necessary modifications in some cases more easily (Section 6.4). Examples of 'non-biotechnological' synthesis are provided by the industrial production of glutamic acid and lysine, conducted on a large scale (several thousand tons per year). DL-Glutamic acid is obtained from acrylonitrile, electrochemical reductive dimerisation and functional group modifications giving the DL compound. DL-Lysine is obtained from caprolactam, through its 3-amino-derivative, which is resolved (Scheme 6.6) with L-pyroglutamic acid before ring-opening to give L-lysine.

6.4 Synthesis of amino acids starting from coded amino acids other than glycine

With the easy availability of many of the natural amino acids, some general methods for the synthesis of more complex structures are based on the modification of simple natural amino acids. An important benefit from this approach is the fact that homo-chirality at the α-carbon atom can be preserved in reactions at side-chains that are in current use.

Thus, D- or L-serine can be converted through the Mitsunobu reaction into the homochiral α-amino-β-lactone, a chiral synthon amenable to ring-opening by organometallic reagents (Pansare and Vederas, 1989) to give β-substituted alanines (Scheme 6.1). β-Iodo-L-alanine (also obtained from L-serine) can be elaborated similarly into the general class of β-substituted alanines (L-serine $\rightarrow$ $H_3N^+CH(CH_2I)CO_2^- \rightarrow H_3N^+CH(CH_2R)CO_2^-$ (Jackson et al., 1989)). L-Aspartic acid and L-glutamic acid serve the same function, electrophiles being substituted at the carbon atom next to the side-chain carboxy group after its deprotonation with lithium di-isopropylamide (Baldwin et al., 1989). As shown in this composite example from a number of research papers, the side-chain carboxy group can be transformed into other functional groups, when one starts with suitably protected glutamates and aspartates (Scheme 6.2).

There are numerous other isolated examples of the conversion of a coded amino acid into another amino acid. These usually amount to applications of straight-

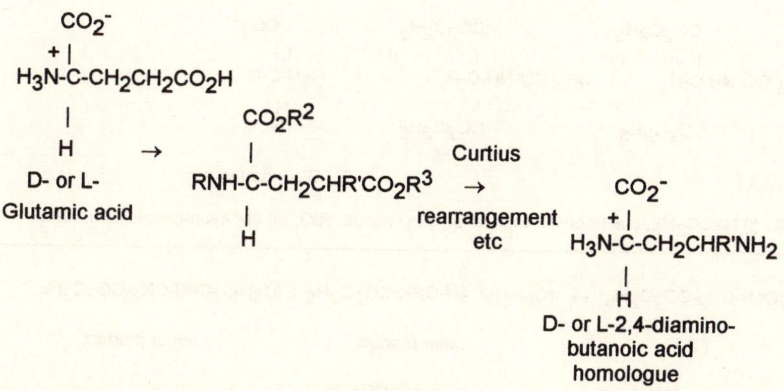

$$
\begin{array}{c}
\text{CO}_2^- \\
\overset{+}{|} \\
\text{H}_3\text{N-C-CH}_2\text{CH}_2\text{CO}_2\text{H} \\
| \\
\text{H} \\
\text{D- or L-} \\
\text{Glutamic acid}
\end{array}
\quad \rightarrow \quad
\begin{array}{c}
\text{CO}_2\text{R}^2 \\
| \\
\text{RNH-C-CH}_2\text{CHR'CO}_2\text{R}^3 \\
| \\
\text{H}
\end{array}
\quad
\begin{array}{c}
\text{Curtius} \\
\rightarrow \\
\text{rearrangement} \\
\text{etc}
\end{array}
\quad
\begin{array}{c}
\text{CO}_2^- \\
\overset{+}{|} \\
\text{H}_3\text{N-C-CH}_2\text{CHR'NH}_2 \\
| \\
\text{H} \\
\text{D- or L-2,4-diamino-} \\
\text{butanoic acid} \\
\text{homologue}
\end{array}
$$

Scheme 6.2.

forward functional group chemistry – e.g. aromatic substitution reactions of phenyl-alanine and tyrosine – that have as their only additional requirement that protection of amino and carboxy groups may need to be considered.

6.5 General methods of synthesis of amino acids starting with a glycine derivative

Simplest of all the laboratory methods, in concept, are those general methods based on the alkylation of glycine derivatives shown in Scheme 6.3, particularly 2-acyl-amidomalonate esters (**1**), Schiff bases (**2**), oxazol-5(4H)-ones (*alias* 'azlactones', **3**) and piperazin-2,5-diones (**4**).

6.6 Other general methods of amino-acid synthesis

The α-amino-acid grouping, —NH—CHR—CO—O—, can be built up from its components through the Strecker synthesis (Equation (6.1) in Scheme 6.4) or by the Bucherer–Bergs synthesis (*alias* hydantoin synthesis; Equation (6.2) in Scheme 6.4). Three general methods – *the diethyl acetamidomalonate, Strecker and Bucherer–Bergs syntheses* – remain the most-used general methods, together with *the oxazolone route* (*the Erlenmeyer 'azlactone' synthesis* shown in Scheme 6.3). An even simpler synthesis, the Miller–Urey experiment in which some of the presumed atmospheric components in pre-biotic eras were shown to combine (Equation (6.3)), is not of practical interest since it gives mixtures with low yields and it cannot be directed towards a predominant target amino acid.

Further general syntheses are shown in Scheme 6.5 (amination of halogenoalka-noic acid derivatives (Equation 6.4), carboxylation or carbonylation of alkylamines (Equation 6.5) and the Ugi 'four-component condensation' (Equation 6.6)). These are useful methods capable of development in certain cases for large-scale syntheses of simple amino acids.

123

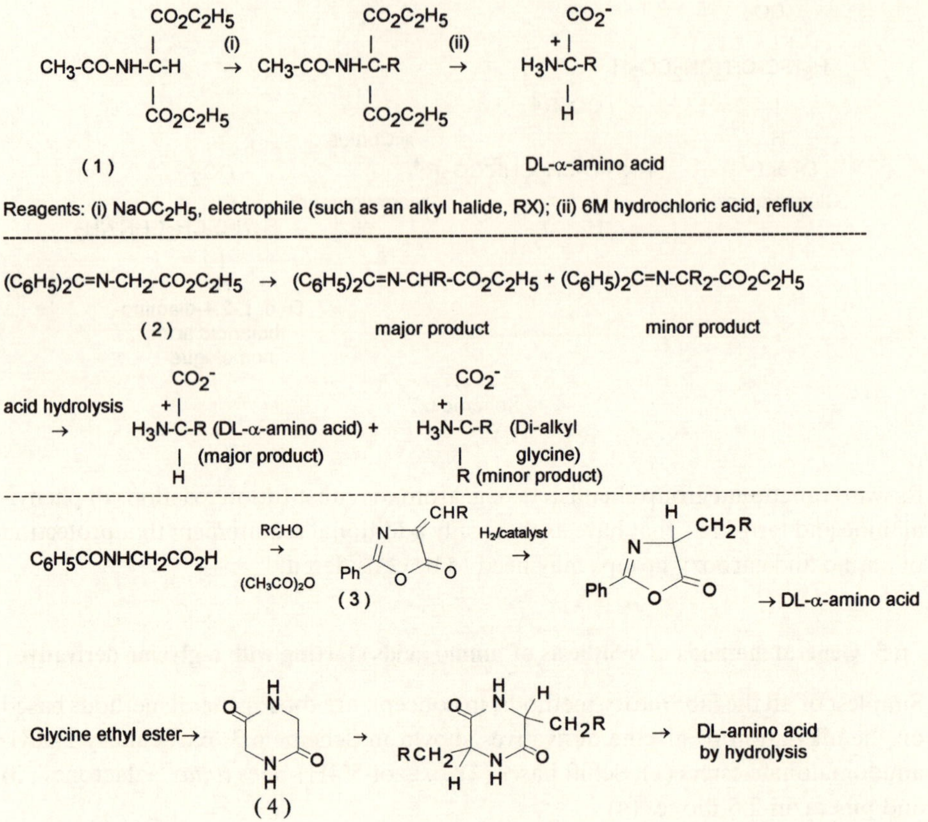

$$
\begin{array}{ccc}
\text{CO}_2\text{C}_2\text{H}_5 & \text{CO}_2\text{C}_2\text{H}_5 & \text{CO}_2^- \\
| & | & +| \\
\text{CH}_3\text{-CO-NH-C-H} \xrightarrow{\text{(i)}} & \text{CH}_3\text{-CO-NH-C-R} \xrightarrow{\text{(ii)}} & \text{H}_3\text{N-C-R} \\
| & | & | \\
\text{CO}_2\text{C}_2\text{H}_5 & \text{CO}_2\text{C}_2\text{H}_5 & \text{H}
\end{array}
$$

(1)   DL-α-amino acid

Reagents: (i) $NaOC_2H_5$, electrophile (such as an alkyl halide, RX); (ii) 6M hydrochloric acid, reflux

----

$(C_6H_5)_2C=N-CH_2-CO_2C_2H_5 \rightarrow (C_6H_5)_2C=N-CHR-CO_2C_2H_5 + (C_6H_5)_2C=N-CR_2-CO_2C_2H_5$

(2)   major product   minor product

$$
\begin{array}{cc}
\text{CO}_2^- & \text{CO}_2^- \\
\text{acid hydrolysis} \quad +| & +| \\
\rightarrow \quad \text{H}_3\text{N-C-R (DL-}\alpha\text{-amino acid) +} & \text{H}_3\text{N-C-R (Di-alkyl} \\
| \text{ (major product)} & | \quad \text{glycine)} \\
\text{H} & \text{R (minor product)}
\end{array}
$$

$C_6H_5CONHCH_2CO_2H$ $\xrightarrow[\text{(CH}_3\text{CO)}_2\text{O}]{\text{RCHO}}$ (3) $\xrightarrow{\text{H}_2/\text{catalyst}}$ $\rightarrow$ DL-α-amino acid

Glycine ethyl ester→   →   → DL-amino acid by hydrolysis

(4)

Scheme 6.3.

Strecker Synthesis

$RCHO + NH_4CN \rightarrow NH_2\text{-CHR-CN} \rightarrow$ DL-α-amino acid   (6.1)

Bucherer-Bergs Synthesis

$$
\begin{array}{ccc}
RCHO + (NH_4)_2CO_3 \rightarrow & NH\text{ - CHR} & \rightarrow \quad \text{DL-}\alpha\text{-amino acid} \\
& | \qquad | & \\
& O=C \quad C=O & \\
& \backslash \quad / & \\
& NH &
\end{array}
$$
(6.2)

Scheme 6.4.

$$N_2 + CH_4 + H_2O \xrightarrow{\text{energy source}} \text{Mixture of amino acids (mostly } \alpha\text{-amino acids)}$$ (6.3)

Gabriel synthesis:

Phthalimide + BrCHRCO$_2$H → DL-α- PhthNCHRCO$_2$H → (hydrazine) amino acid   (6.4)

(There are several similar routes aimed at introducing an amino group, e.g. reaction of NaN$_3$
  with an α-halogeno-acid, followed by reduction of the azido-function)

Curtius rearrangement:

```
   CH3 Ph        CH3 Ph        CH3 Ph       CH3 CO2H
    \ /           \ /           \ /    [O]   \ /
     C      →      C      →      C      →     C
    / \           / \           / \    H2O   / \
  HO2C  H        H2N  H      PhCONH  H      H2N  H
```

R(-)-phenylpropanoic acid   R(-)-alanine

(Several similar rearrangements, e.g the **Schmidt rearrangement**, can be used analogously,
  to introduce the amino function)

Biomimetic Amination of an α-Keto-acid:

O=CR^1-CO$_2$H + pyridoxal-5'-phosphate + NH$_3$ → NH$_2$CHR^{1}CO$_2$H

Amination at the α-position of an alkanoic acid:

```
          LDA           BocN=NBoc   BocN-NHBoc  1) TFA
RCHCO2R'   →   RCH⁻CO2R'     →          |          →   α-amino-acid
                                     RCHCO2R'  2) H2/Ni
```

Amidocarbonylation of an aldehyde or a ketone:

R^{1}CONH$_2$ + R^{2}CHO + CO → R^{1}CONHCHR^{2}CO$_2$H   (6.5)

(catalyst)

Ugi "Four-Component Condensation":

R^{1}NC + R^{2}R^{3}CO + R^{4}NH$_2$ + R^{5}CO$_2$H → R^{5}CONR^{4}CR^{3}R^{2}CONHR^1 (6.6)

Scheme 6.5.

6.7 Resolution of DL-amino acids

The requirements for homochirally pure α-amino acids have not ruled out any of
these general synthetic methods (which all give racemic products), since *resolution*
of DL-α-amino acids and their derivatives is a simple, albeit time-consuming, solution to this need. Classical methods for resolution include physical separation of the
DL-amino acids themselves (by chromatography on a chiral phase; e.g. resolution of
DL-tryptophan over cellulose, see Section 4.15), fractional crystallisation of certain
racemates or supersaturated solutions (through seeding with crystals of one enan-

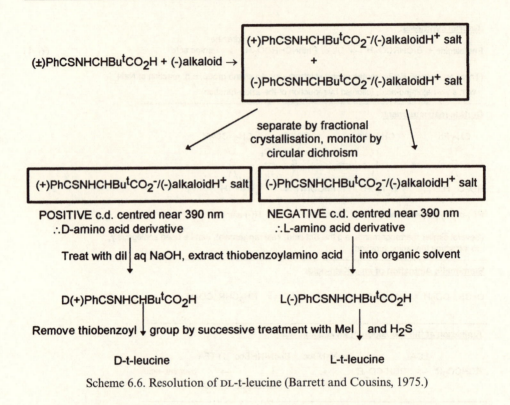

Scheme 6.6. Resolution of DL-t-leucine (Barrett and Cousins, 1975.)

tiomer) and, more commonly, separation by crystallisation of diastereoisomeric derivatives (alkaloid salts of N-acylated DL-amino acids; fractional crystallisation of DL-amino acids derivatised with homochiral N-acyl and/or O-alkyl ester groups). Scheme 6.6 displays a typical amino-acid-resolution procedure applicable both on the laboratory scale and industrially (e.g. L-lysine manufacture, Section 6.3.2).

Enzymic resolution is also generally useful. At first sight it is of restricted applicability, since most of the classical methods are based on the selectivity of a proteinase for catalysing the hydrolysis of the L enantiomer of an N-acyl derivative of a DL-amino acid (Equation (6.7)) or of a DL-amino acid ester. The normal substrates for these enzymes are derivatives of particular coded amino acids.

$$\text{DL-R}^1\text{CONHCHR}^2\text{CO}_2\text{H} \xrightarrow{\text{trypsin}} \text{D-R}^1\text{CONHCHR}^2\text{CO}_2\text{H} + \text{L-NH}_2\text{CHR}^2\text{CO}_2\text{H} \qquad (6.7)$$

However, the range of types of amino acids that can be resolved in this way is much greater than just the natural substrates (i.e. peptides made up of the twenty coded amino acids), because methods to relax the specificity of the enzymes have been found, in some cases by using organic solvents for the reactions. Penicillin acylase from *Escherichia coli* and an aminoacylase from *Streptovercillium olivoreti-*

126

culi have been used for the preparative-scale resolution of phenylalanines and phenylglycines carrying fluoro-substituents in the benzene ring (Kukhar and Soloshonok, 1995; Soloshonok *et al.*, 1993).

The use of enzymes with hydantoins (Equation (6.2) in Scheme 6.4) is particularly suitable and can be quite simple since various bacteria possess D-hydantoinase activity and can be used conveniently in a 'whole-cells' procedure that avoids the need to extract and purify the actual enzymes concerned. As in the principle shown in the 'trypsin' equation just above, one hydantoin is converted through hydrolysis into the D-amino acid, whereas the other remains unaffected.

6.8 Asymmetric synthesis of amino acids

The correct usage of the term asymmetric synthesis implies the involvement of at least one stereoselective reaction for the preferential or exclusive generation of one particular configuration at the chiral centre in the amino acid that emerges at the end of the synthesis (Barrett, 1985; Williams, 1989). The general methods of amino-acid synthesis discussed above can all, in principle, be carried out in the stereo-selective mode, but then depend for their enantioselectivity on the use of *a chiral catalyst* or on the presence of *a chiral centre in the ester moiety* of the glycine syn-thons. The use of a chiral catalyst (such as a *Cinchona* alkaloid) is illustrated in the phase-transfer alkylation of imines (**2** in Scheme 6.3), giving better than 99% enan-tiomeric excess when the alkylating agent is 4-chlorobenzyl chloride in the synthesis of 4-chloro-L-phenylalanine (O'Donnell and Wu, 1989).

The approach exploiting a chiral centre that is *already in the synthon* is effective in a number of cases. The chiral moiety in the synthon diverts a reaction at a nearby prochiral centre in favour of one enantiomer (*asymmetric induction*). An excellent example of the latter is the Schöllkopf method (**4** in Scheme 6.3, see also **5** in Scheme 6.7); hydrogenation of 'azlactones' (**3** in Scheme 6.3) using a homogeneous chiral catalyst is one route illustrating the former approach. Use of chiral five-membered heterocyclic compounds (e.g., **6** and **7**) offers an alternative successful approach to asymmetric amino-acid synthesis.

In many of these cases, the new chiral centre is generated in an achiral starting material (e.g., the oxazolone), whereas in others (e.g., the imidazolidinone) the start-ing compound is homochiral and cannot be recovered. However, the 'chiral auxil-iary' approach in which a homochiral reactant is recovered unchanged at the end of an asymmetric synthesis is illustrated in some of the examples in Scheme 6.7 (the Belokon and oxazolidinone methods are good examples). Many recent syntheses have used all these methods and close variants thereof.

To some extent, it is a matter of perceived ease of working, or favourable econom-ics, when it comes to choice of method; the piperazinedione route can be operated on a scale of several hundreds of grams (Schöllkopf *et al.*, 1985). Nonetheless, a major consideration is the stereochemical efficiency that is involved (i.e. the

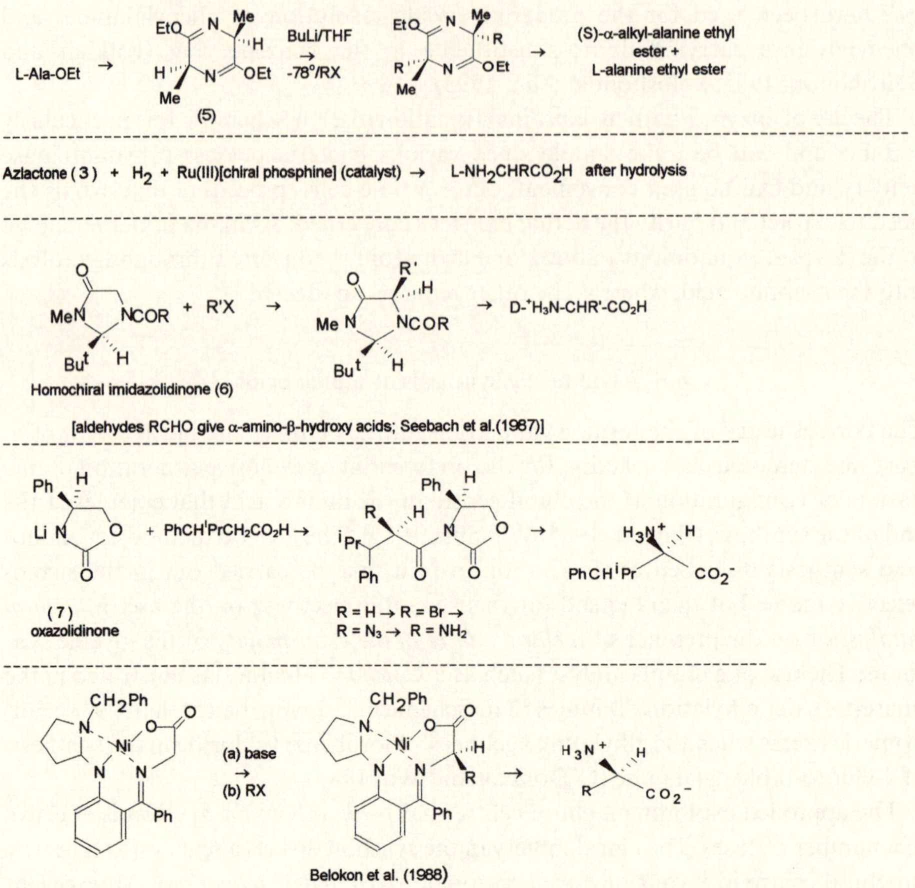

L-Ala-OEt →

BuLi/THF
→
-78°/RX

→ (S)-α-alkyl-alanine ethyl ester + L-alanine ethyl ester

(5)

Azlactone (3) + H₂ + Ru(III)[chiral phosphine] (catalyst) → L-NH₂CHRCO₂H after hydrolysis

+ R'X →

→ D-⁺H₃N-CHR'-CO₂H

Homochiral imidazolidinone (6)

[aldehydes RCHO give α-amino-β-hydroxy acids; Seebach et al.(1987)]

+ PhCH'PrCH₂CO₂H →

→

(7)
oxazolidinone

R = H → R = Br →
R = N₃ → R = NH₂

(a) base
→
(b) RX

→

Belokon et al. (1988)

Scheme 6.7.

diastereoisomer excess involved when one starts with a homochiral auxiliary), since a more difficult purification of the product to complete enantiomeric purity is involved when small enantiomer excesses are achieved.

In the Schöllkopf piperazinedione method, namely alkylation of the 2,5-diethoxy compound prepared from L-alanine methyl ester, values greater than 90% are routinely achieved for the alkylation yield and for the diastereoisomeric excess of the product (Allen et al., 1992). Similar results have been reported for the Belokon method and for the Seebach imidazolidinone method (though there are rather low alkylation yields in some cases).

6.9 References

General sources of information on general synthetic methods for amino acids (Barrett, 1985) and on asymmetric synthesis (Williams, 1989) are listed in the Foreword.

Allen, M. S., Hamaker, L. K., La Loggia, A. J. and Cook, M. J. (1992) *Synth. Commun.*, **22**, 2077.

Baldwin, J. E., North, M., Flinn, A. and Moloney, M. G. (1989) *Tetrahedron*, **45**, 1453.

Barrett, G. C. and Cousins, P. R. (1975) *J. Chem. Soc., Perkin Trans. I*, 2313.

Belokon, Y. N., Sagyan, A. S., Djamgaryan, S. M., Bakhmutov, V. I. and Belikov, V. M. (1988) *Tetrahedron*, **44**, 5507.

Chattopadhyay, S. and Das, A. (1990) *FEMS Microbiol. Lett.*, **72**, 195.

Eggeling, L. (1994) *Amino Acids*, **6**, 261.

Goldberg, I. and Williams, R. A. (1991) *Biotechnology of Food Ingredients*, Van Nostrand-Reinhold, New York.

Greenstein, J. P. and Winitz, M. (1961) *Chemistry of the Amino Acids*, Wiley, New York.

Kukhar, V. P. and Soloshonok, V. A. (Ed.) (1995) *Fluorine-Containing Amino Acids: Synthesis and Properties*, Wiley, Chichester.

Jackson, R. F. W., James, K., Wythes, M. J. and Wood, A. (1989) *J. Chem. Soc., Chem. Commun.*, 644.

O'Donnell, M. J. and Wu, S. (1989) *J. Amer. Chem. Soc.*, **111**, 2353.

Pansare, S. V. and Vederas, J. C. (1989) *J. Org. Chem.*, **54**, 2311.

Rozzell, J. D. and Wagner, F. (1992) *Biocatalytic Production of Amino Acids and Their Derivatives*, Wiley, New York.

Schöllkopf, U., Lonsky, R. and Lehr, P. (1985) *Liebigs Ann. Chem.*, 413.

Seebach, D., Juaristi, E., Miller, D. D., Schickli, C. and Weber, T. (1987) *Helv. Chim. Acta*, **70**, 237.

Soloshonok, V. A., Galaev, I. Y., Svedas, V. K., Kozlova, E. V., Kotif, N. V., Shishkina, I. P., Galushko, S. V., Rozhenko, A. B. and Kukhar, V. P. (1993) *Bioorg. Khim.*, **19**, 467.

Wichers, H. J., Malingre, T. M. and Huizing, H. J. (1985) *Planta*, **166**, 421.

7

Methods for the synthesis
of peptides

7.1 Basic principles of peptide synthesis and strategy

The synthesis of a dipeptide, $NH_3^+ CHR^1CONHCHR^2COO^-$, from the constituent amino acids involves forming the peptide bond so that the amino-acid sequence is correct and enantiomerisation (Section 7.7) at the chiral α-carbon atoms is avoided. The latter point does not arise, of course, with glycine. In order to produce the correct sequence and to prevent the formation of a mixture of higher peptides, the amino group of the intended *N*-terminal residue and the carboxy group of the intended *C*-terminal residue are normally protected.

In the synthesis of higher peptides, the polypeptide chain can be built up one unit at a time in either direction. For example, an octapeptide could be synthesised in several stages proceeding through a dipeptide, a tripeptide, a tetrapeptide and so on. Alternatively, the octapeptide could be formed from two tetrapeptide units (fragment condensation), which in turn might be built up one unit at a time or formed from two dipeptides. Apart from chemical considerations, the overall yield would depend on the route selected. For example, if the synthesis of each peptide bond could be achieved with a yield of 80%, the stepwise procedure of adding one amino-acid residue at a time would give an overall yield of 21% relative to the first two amino acids used. In contrast, the synthesis which proceeds through four dipeptides to two tetrapeptides and thence to the octapeptide would give an overall yield of 51%. On the face of it, the last method appears to be the best, yet it is the first route that is most commonly used because other factors must be considered. For example, the risk of enantiomerisation varies with the synthetic route selected (Section 7.7). It is advantageous to have Gly or Pro as the *C*-terminal residue of a protected peptide whose carboxy group is to be linked to the amino group of another peptide derivative in the fragment condensation, since the risk of enantiomerisation is eliminated for *C*-terminal Gly and considerably reduced for *C*-terminal Pro (Section 7.7). Moreover, in the solid-phase method of peptide synthesis in which one residue

130

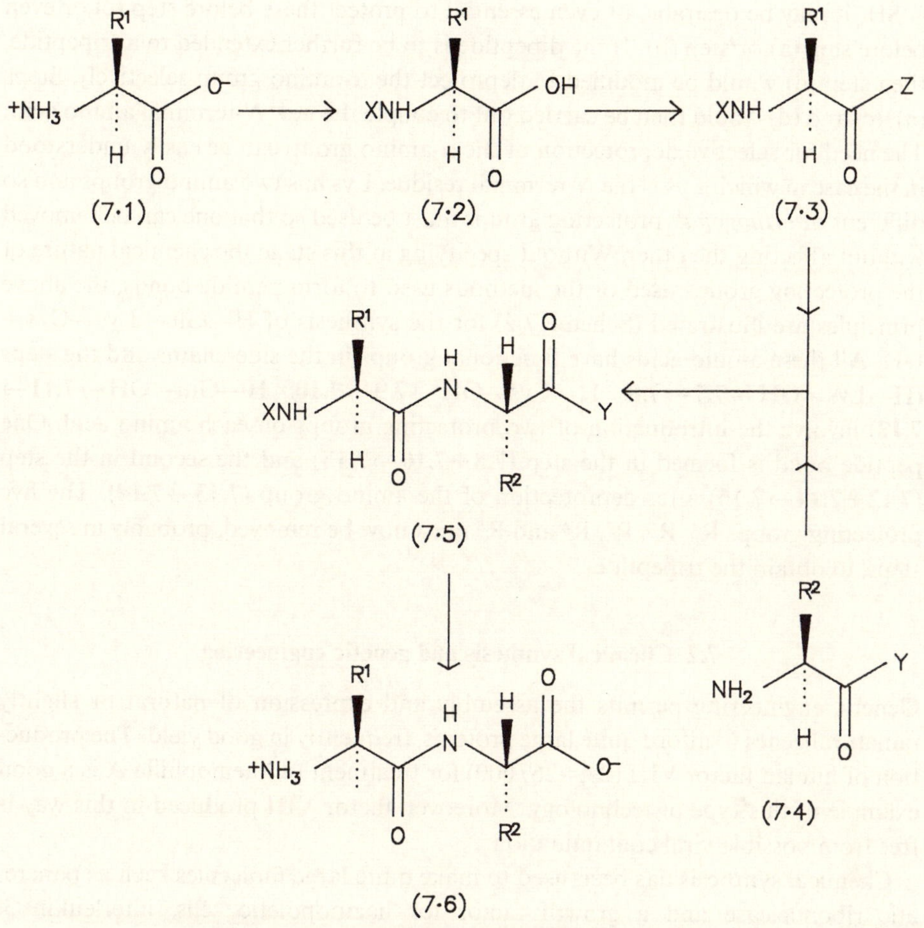

Scheme 7.1.

is coupled at a time (Section 7.9), it is now usually possible to achieve yields better than 99%. With this performance, the yield of an octapeptide would be over 93%. It is important to note that, although the biosynthesis of proteins involves the addition of one residue at a time from the *N*- to the *C*-terminus, chemical synthesis is always carried out in the opposite direction (Section 7.7).

The synthesis of a dipeptide in general involves four steps (Scheme 7.1): (a) protection of the amino group of the amino acid that is to be the *N*-terminal residue (**7.1→7.2**), (b) protection of the carboxy group of the amino acid that is to be the *C*-terminal residue in the dipeptide (**7.4**), (c) activation of the carboxy group of the *N*-terminal amino acid (**7.2→7.3**) and formation of the peptide bond to give a protected dipeptide (**7.3+7.4→7.5**) and (d) removal of protecting groups (**7.5→7.6**). If the amino acids contain functional groups such as —NH₂, —COOH, —OH, and

131

—SH, it may be desirable or even essential to protect these before step (c) or even before step (a) or step (b). If the dipeptide is to be further extended to a tripeptide, then step (d) would be modified to deprotect the α-amino group selectively. Steps (a), (c) and (d) would then be carried out to couple the new N-terminal amino acid. The need for selective deprotection of the α-amino group can be easily understood in the case in which Lys is the N-terminal residue. Lys has two amino groups and so different or *orthogonal*[1] protecting groups must be used so that one can be removed without affecting the other. Without specifying at this stage the chemical nature of the protecting groups used or the methods used to form peptide bonds, the above principles are illustrated (Scheme 7.2) for the synthesis of H—Glu—Lys—Cys—OH. All these amino acids have functional groups in the side-chains and the steps (H—Lys—OH→**7.7**→**7.8**, H—Cys—OH→**7.9**→**7.10**, H—Glu—OH→**7.11**→**7.12**) involve the introduction of two protecting groups on each amino acid. One peptide bond is formed in the step (**7.8**+**7.10**→**7.13**) and the second in the step (**7.12**+**7.14**→**7.15**) after deprotection of the amino group (**7.13**→**7.14**). The five protecting groups R^1, R^2, R^3, R^4 and R^5 must now be removed, probably in several steps, to obtain the tripeptide.

7.2 Chemical synthesis and genetic engineering

Genetic engineering permits the assembly and expression of natural or slightly unnatural genes to afford quite large proteins, frequently in good yield. The production of human factor VIII ($M_r \approx 267\,000$) for treatment of haemophilia A is a good example of this type of technology. Moreover, factor VIII produced in this way is free from possible viral contamination.

Chemical synthesis has been used to make quite large molecules such as pancreatic ribonuclease and a growth factor for haemopoietic cells, interleukin 3. Generally, however, chemical synthesis is used for the synthesis of polypeptides containing up to about 120 amino residues. Most polypeptide hormones, but not all enzymes, are thus accessible by chemical synthesis. Genetic engineering is preferred for the synthesis of larger polypeptides. It is important to stress that chemical synthesis possesses some advantages over genetic engineering since the latter is mainly applicable to only the twenty amino acids defined by the genetic code. Some recent experiments have permitted the introduction of a single unnatural amino acid, but this is a long way short of the versatility of chemical synthesis. Design of peptides for drug use usually requires the incorporation of some unnatural amino acids, including surrogate peptide bonds, to achieve molecular longevity or selectivity of

---

[1] *Orthogonal* normally means 'right-angled or situated at right-angles'. In the context of peptide synthesis, the term 'orthogonal' has nothing to do with the absolute or relative geometries of protecting groups. Rather, it is best to think of orthogonal groups as having vectors at right-angles that represent the ease of deprotection by particular reagents. Thus, one of two orthogonal groups will be completely removed by one reagent whereas the other orthogonal group will remain unaffected by this reagent.

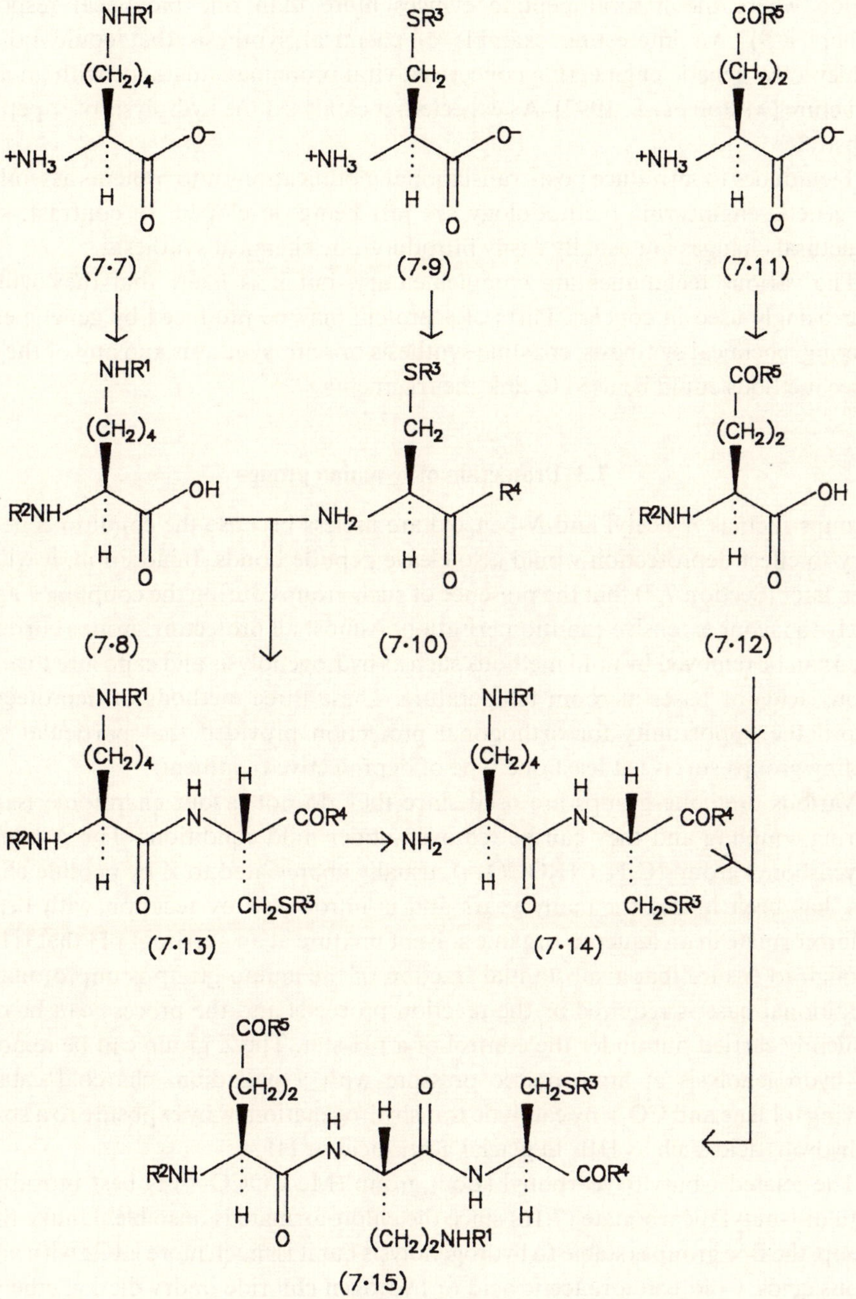

Scheme 7.2.

action where the natural peptide evinces more than one biological response (Chapter 9). An interesting example of chemical synthesis that could not be achieved by genetic engineering concerns a viral proteinase analogue with an all-D structure (Milton *et al.*, 1992). As expected, it catalysed the hydrolysis of D-peptide substrates.

Techniques to introduce post-translational modifications into proteins assembled by genetic engineering methodology are just being developed. In contrast, such structural changes are usually easily introduced by chemical synthesis.

The various techniques are complementary, but it is likely that they will be increasingly used in concert. Parts of a protein may be produced by genetic engineering, chemical synthesis, enzymic synthesis or semi-synthesis and any of the last three methods could be used to link the fragments.

7.3 Protection of α-amino groups

Groups such as *N*-acetyl and *N*-benzoyl are useless because the conditions necessary to effect deprotection would also cleave peptide bonds. In addition, it will be seen later (Section 7.7) that the presence of such groups during the coupling stage is likely to favour extensive enantiomerisation. Almost all protecting groups currently used can be removed by mild methods such as hydrogenolysis and exposure to anhydrous acids or bases at room temperature. These three methods of deprotection afford the opportunity for orthogonal protection provided that particular protecting groups survive at least one type of deprotective treatment.

Various urethane groups are used since they do not favour enantiomerisation during coupling and they can be removed under mild conditions. The *N*-benzyloxycarbonyl group ($C_6H_5CH_2OCO$—), usually abbreviated to Z by peptide chemists, has been in use for many years and is introduced by reaction with benzyl chloroformate in an aqueous organic solvent mixture at an apparent pH that is high enough to ensure that a substantial fraction of the amino group is unprotonated. Additional base is required as the reaction proceeds and the process can be conveniently carried out under the control of a pH-stat. The Z group can be removed by hydrogenolysis at atmospheric pressure with a palladium charcoal catalyst (giving toluene and CO_2), by catalytic transhydrogenation or by exposure to a strong anhydrous acid such as HBr in glacial acetic acid or HF.

The related t-butyloxycarbonyl (Boc) group (Me_3COCO—) is best introduced with di-t-butyl dicarbonate (**7.16**) since the chloroformate is unstable. Unlike the Z group, the Boc group is stable to hydrogenolysis but it is much more labile with anhydrous acids. Cold trifluoroacetic acid or hydrogen chloride in dry diethyl ether are convenient deprotecting agents, but the transient formation of the t-butyl carbocation can lead to unwanted alkylation of susceptible groups such as the indole ring of Trp and the thioether function of Met during deprotection of the peptide. It is common practice to add a suitable compound to act as a decoy for alkylation by

$$CH_3 - \underset{\underset{CH_3}{|}}{\overset{\overset{CH_3}{|}}{C}} - O - \underset{O}{\overset{\|}{C}} - O - \underset{O}{\overset{\|}{C}} - O - \underset{\underset{CH_3}{|}}{\overset{\overset{CH_3}{|}}{C}} - CH_3$$

(7·16)

carbocations. Thiophenols and thioethers are commonly used during removal of Boc and related groups.

The 9-N-fluorenylmethoxycarbonyl (Fmoc) group, which can be introduced (Scheme 7.3) using the corresponding chloroformate (7.17), is a third type of urethane protecting group. It is stable to the acidolytic conditions that remove Z and Boc groups, but it is very labile to bases such as piperidine and morpholine. After peptide synthesis (7.18→7.19) base-catalysed removal of a proton from C_9 of the fluorene nucleus causes an elimination reaction with the formation of dibenzofulvene (7.20) and the free amino compound. The dibenzofulvene either polymerises or adds excess amine. The Fmoc group can also be slowly removed by hydrogenolysis, but this method is not usually chosen.

A fourth urethane protecting group, the N-allyloxycarbonyl group (Alloc) is introduced in the usual way using allyl chloroformate or diallyl dicarbonate. Its main interest concerns its removal by a Pd-catalysed hydrostannolysis with tributyltin hydride (Scheme 7.4). It thus provides orthogonal protection without the need to expose the peptide to acid, conditions that would cleave, for example, O-glycoside derivatives of peptides.

The 2-(4-biphenylyl)propyl-2-oxycarbonyl (Bpoc) group (7.21) is even more sensitive than the Boc group towards acidolytic cleavage and you should be able to explain why this is so. This protecting group enjoyed a period of popularity mainly because the mild conditions required to remove it leave other groups such as t-butyl esters unaffected. It is an expensive group to use and has largely gone out of fashion. Similarly, the related 1-adamantyloxycarbonyl group (7.22) has also passed its peak of popularity, although it is still occasionally used to protect the guanidino group of arginine (Section 7.5). Many other reagents have been suggested for protecting the α-amino group but are seldom used (Jones, 1994).

7.4 Protection of carboxy groups

As will be seen later, peptide-bond formation can be achieved by converting the intended N-protected N-terminal residue into a reactive derivative of the carboxy function in the absence of the C-terminal moiety to which it is to be coupled. In these circumstances, it is not always necessary to protect the carboxy group of the latter. The C-terminal moiety then has both the α-amino group and the terminal

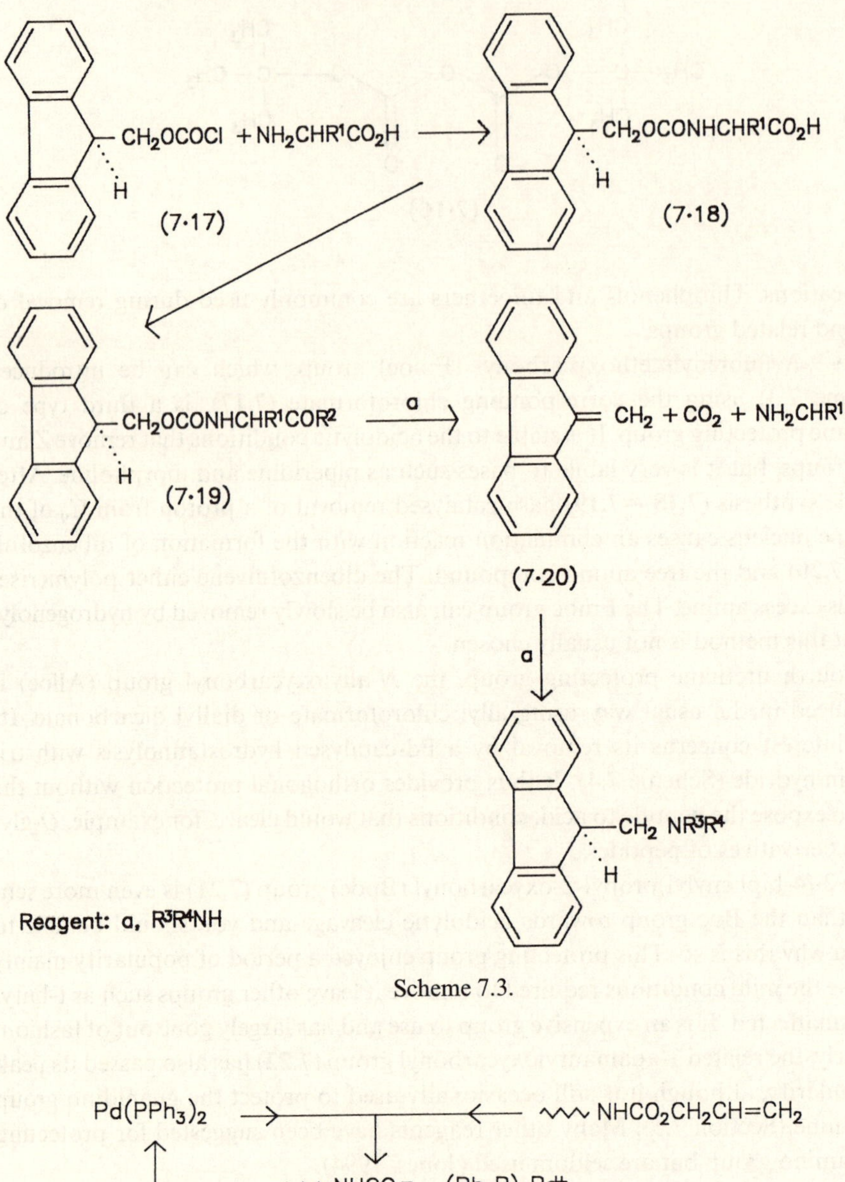

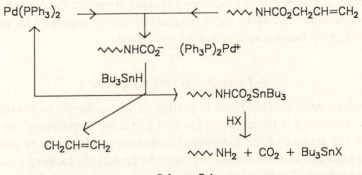

Reagent: a, R³R⁴NH

Scheme 7.3.

Scheme 7.4.

(7·21) (7·22)

carboxy group free and its dipolar ion character (Chapter 3) will normally require the coupling step to be carried out in aqueous solution. This in turn requires that the reactive derivative of the *N*-protected amino acid to which it is to be coupled must be much more stable to hydrolysis than it is to nucleophilic attack by the amino group. The coupling must also be carried out at a pH high enough to ensure that a substantial fraction of the amino component is in the unprotonated form. This increases the danger that the reactive derivative of the *N*-protected amino acid will suffer hydrolysis rather than coupling. In spite of these rather restrictive conditions, numerous peptides have been synthesised without protecting the carboxy group, especially when the *N*-protected amino acid is introduced in the form of a reactive ester (Section 7.8). Although the omission of carboxy protection saves two steps, the increasing use of solid-phase synthesis of peptides (Section 7.9) has diminished the importance of this approach, except for the synthesis of short peptides.

More generally, the *C*-terminal group is protected as an ester (Jones, 1994). Since esters of amino acids and peptides do not have a dipolar ion structure, they are soluble in aprotic solvents. There is also a striking difference in the pK_a values of an amino ester and its esters (Chapter 3). That the pK_a values of the esters are lower means that the NH_3^+ group can lose its proton to a weaker base. These factors together mean that peptide-bond formation can be carried out in aprotic solvents with less risk of the reactive derivative of the *N*-protected *N*-terminal amino acid being racemised. It should also be noted that the lower pK_a of the amino group of an amino-acid ester implies a weaker nucleophilic character. This is seldom a significant factor in peptide synthesis.

Methyl and ethyl esters enjoyed a long period of popularity because they can readily be prepared as crystalline hydrochlorides after allowing a solution of the amino acid in methanolic or ethanolic HCl to stand overnight or by reaction with a mixture of thionyl chloride and the appropriate alcohol. The salts can be converted into the free bases by shaking them briefly with a solution of ammonia in chloroform, filtering and evaporating the filtrate under reduced pressure. Unfortunately, the removal of methyl or ethyl ester protecting groups at the end of a peptide synthesis usually requires the use of alkali and this can cause enantiomerisation (Section 7.7).

Benzyl esters and substituted benzyl esters are also easily prepared but possess the advantage that deprotection can be achieved without using alkali. Benzyl esters are

CH$_2$COOCMe$_3$ OH$^-$ →   CH$_2$ —— CO

|

−NHCHCONHR

(7·23)

CH$_2$ —— CO
         NHR

|

−NHCH —— CO

(7·24)

cleaved by hydrogenolysis using a palladium–charcoal catalyst. The presence of divalent sulphur derivatives such as Met and Cys residues can inhibit deprotection. Alternatively, benzyl esters are cleaved by strong, anhydrous acids such as hydrogen bromide in acetic acid. The solid-phase method of synthesis usually involves the attachment of the growing peptide chain to the matrix through a benzyl ester group (Section 7.9). Allyl esters are also used and these are deprotected by palladium-catalysed hydrostannolysis with Bu$_3$SnH, analogously to the removal of Alloc groups described above.

A common orthogonal method of protecting carboxy groups uses the t-butyl group. These esters are usually prepared from amino acids or their *N*-protected derivatives by treatment with isobutene in the presence of a strong acid such as sulphuric acid or toluene-*p*-sulphonic acid. Treatment with hydrogen chloride in organic solvents or with trifluoroacetic acid at room temperature effects deprotection. This ready removal of the t-butyl ester group makes it very suitable for the protection of the side-chain carboxy groups of Asp and Glu residues. The t-butyl ester group is stable to alkaline hydrolysis and hydrogenolysis. Occasionally, t-butyl and other alkyl esters of Asp can undergo cyclisation under basic conditions (**7.23** →**7.24**), reinforcing the comments about the undesirability of using alkali for deprotection.

7.5 Protection of functional side-chains

7.5.1 *Protection of ε-amino groups*

The need to use orthogonal protection on the α- and ε-amino groups of lysine has been explained above (Section 7.1) and it only remains necessary to describe how these groups are introduced. The α-amino and carboxylate groups constitute a bidentate ligand for the Cu^{2+} ion. Reaction of lysine with a chloroformate, for example, in the presence of an excess of Cu^{2+} (Scheme 7.5) permits selective protection of the ε-amino group (**7.25**→**7.26**→**7.27**). It has been found that the Z group is less stable on an ε-amino group than it is on an α-amino group. This slight difficulty can be overcome by using either 2-chlorobenzyloxycarbonyl or 2,6-dichlorobenzyloxycarbonyl groups to protect the ε-amino group.

Protection of ε-amino groups in proteins by means of the trifluoroacetyl group was introduced many years ago in order to limit the action of trypsin to the cleav-

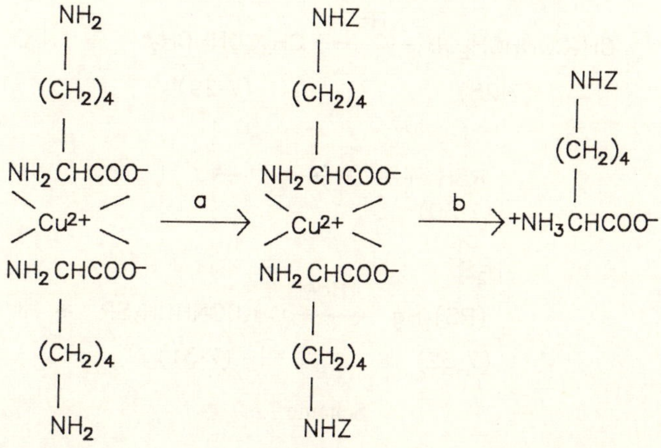

NH$_2$ NHZ

|

(CH$_2$)$_4$ (CH$_2$)$_4$ NHZ

| | |

NH$_2$ CHCOO$^-$ NH$_2$ CHCOO$^-$ (CH$_2$)$_4$

Cu^{2+} $\xrightarrow{a}$ Cu^{2+} $\xrightarrow{b}$ $^+$NH$_3$ CHCOO$^-$

NH$_2$ CHCOO$^-$ NH$_2$ CHCOO$^-$

| |

(CH$_2$)$_4$ (CH$_2$)$_4$

| |

NH$_2$ NHZ

(7·25)   (7·26)   (7·27)

Reagents: a, C$_6$H$_5$CH$_2$OCOCl; b, H$_2$S or EDTA

Scheme 7.5.

age of arginyl peptide bonds in sequencing studies (Chapter 5). The ε-trifluoroacetyl group can be used in peptide synthesis. It is easily removed by exposure to bases such as piperidine but is stable to mild acid treatment. It is thus orthogonal to Z and Boc but not to Fmoc groups. The relatively small trifluoroacetyl group does not seriously decrease the solubility of peptide derivatives, in contrast to bulkier groups, which is sometimes a useful property.

7.5.2 Protection of thiol groups

Since the formation of a peptide bond usually involves nucleophilic attack of an amino group on an activated carboxy group, a potentially strong nucleophile such as the thiol group must be protected (Hiskey, 1981). The earliest technique involved the formation of a benzyl thioether or a substituted benzyl thioether. These derivatives are easily accessible to reaction with a benzyl halide under basic conditions. *S*-Benzyl groups were originally removed with sodium in liquid ammonia, but this method has been superseded by the use of strong acids such as trifluoro-methanesulphonic acid and HF. Optimisation involves finding a group that is sufficiently stable to withstand the reagents required to deprotect blocked α-amino groups and to form new peptide bonds yet sufficiently labile to be removed easily at the end of the synthesis. The *S*-4-methoxybenzyl group is one of the most suitable.

Ironically, a very suitable reagent for protecting the thiol group had been described in 1905 but was not applied to peptide synthesis until 1972 (Veber *et al.*, 1972), a salutary reminder that a lot of useful chemistry may be lurking in the old

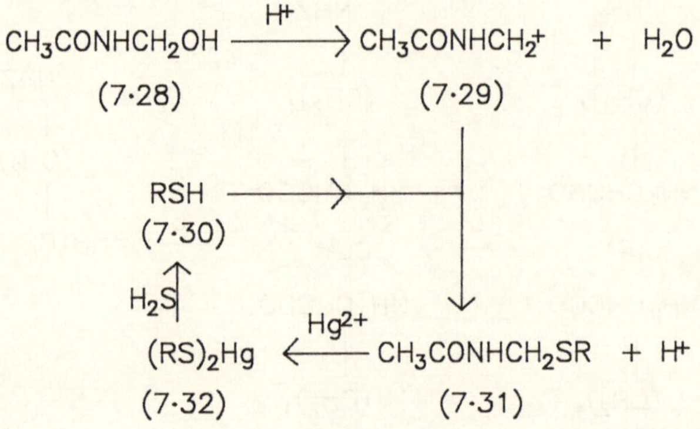

$$CH_3CONHCH_2OH \xrightarrow{\ H^+\ } CH_3CONHCH_2^+ \ + \ H_2O$$

$$(7 \cdot 28) \qquad\qquad\qquad (7 \cdot 29)$$

$$RSH \longrightarrow$$

$$(7 \cdot 30)$$

$$H_2S \uparrow$$

$$(RS)_2Hg \ \xleftarrow{\ Hg^{2+}\ } CH_3CONHCH_2SR \ + \ H^+$$

$$(7 \cdot 32) \qquad\qquad\qquad (7 \cdot 31)$$

Scheme 7.6.

literature awaiting re-discovery and re-deployment. N-Hydroxyacetamide (**7.28**), which is produced by the interaction of acetamide and formaldehyde in the presence of K_2CO_3, reacts with thiol groups such as those in cysteine derivatives (**7.30**) in dilute aqueous acid (**7.28**→**7.29**→**7.31**) (Scheme 7.6). The S-acetamidomethyl group (Acm) is stable against most reagents used in peptide synthesis, but is cleaved by Hg^{2+} ions followed by treatment with H_2S (**7.31**→**7.32**→**7.30**). It is also removed by HF and by treatment with mild oxidising agents, such as by I_2 in MeOH and by $Tl(CF_3COO)_3$. These oxidative methods simultaneously effect the formation of cystine peptides.

7.5.3 Protection of hydroxy groups

Because the hydroxy group is considerably less nucleophilic than the thiol group, the case for its protection in derivatives of serine, threonine and tyrosine is less compelling (Stewart, 1981). Indeed, many peptides containing these amino acids have been synthesised without protection of the hydroxy group. With the current trend towards the synthesis of longer, more valuable peptides, however, a safety-first policy usually prevails. Mention of some possible side reactions will underline the wisdom of this approach. For example, removal of Z groups from peptides bearing unprotected hydroxy groups by treatment with HBr in CH_3CO_2H can lead to O-acetylation. Again, exposure to strong acids can lead to $N \rightarrow O$ migration (**7.33**→ **7.34**). Protection of the hydroxy group of tyrosine is even more important, since addition of base might generate the phenoxide ion, which is a powerful nucleophile.

Protection of hydroxy groups is commonly effected by forming the t-butyl ether by reaction with isobutene in the presence of sulphuric or toluene-p-sulphonic acid, just like with carboxy groups (Section 7.4). Consequently, both groups may be blocked simultaneously by starting with the free acid. t-Butyl ethers are readily cleaved by trifluoroacetic acid.

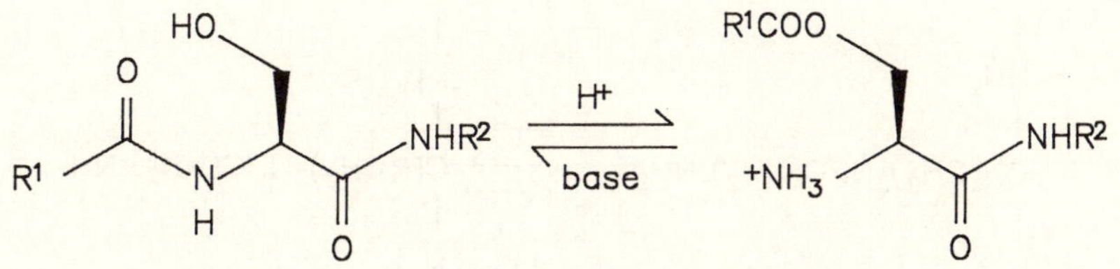

$$(7\cdot33) \qquad\qquad (7\cdot34)$$

Protection of hydroxy groups by benzyl groups is slightly less favoured. First, O-benzyl groups for protection of tyrosine residues are rather too labile with acid. Secondly, when acidolytic cleavage is carried out, $C_6H_5CH_2^+$ ions can effect C-benzylation of the tyrosine side-chain. Addition of a soft nucleophile such as anisole or thioanisole as a competing electrophile scavenger can usually prevent this side reaction. This technique is particularly useful in solid-phase synthetic work, in which detachment of the peptide product from an insoluble support and deprotection are frequently carried out simultaneously. Alternatively, the tyrosine hydroxy group can be protected by the 2-bromobenzyloxycarbonyl (BrZ) group, which is cleanly removed by strong acids.

7.5.4 Protection of the guanidino group of arginine

Although the guanidinium group in the side-chain of arginine has a pK_a of about 13 and is therefore protonated under most conditions, it is generally accepted that protection with a group that will substantially diminish the nucleophilicity of the conjugate base is a sensible precaution in order to prevent lactamisation (**7.35→7.36**) (Scheme 7.7) occurring as an undesired alternative to the formation of a peptide bond.

A great many protecting groups have been examined, including N^G-nitro, N^G,N^G-bisbenzyloxycarbonyl, N^G,N^G-bis-1-adamantyloxycarbonyl and various N^G-arylsulphonyl groups. After application of the usual criteria of ease of introduction, orthogonality to the usual α-N-protecting groups and ease of removal under mild conditions, few groups remain in contention. The most favoured groups are based on the benzenesulphonyl structure. All are removed by strong acids, the severity of conditions required depending on the substituents in the aryl ring. Unfortunately, the cheapest protecting group, toluene-p-sulphonyl, requires HF or CF_3SO_3H in the presence of anisole for its removal. The rationale leading to optimisation of the design of an arylsulphonyl protecting group involved several steps by different workers. The p-methoxybenzenesulphonyl group is more readily removed than is benzenesulphonyl by acid due to the electron-donating property of the p-methoxy group, which promotes the formation of $ArSO_2^+$ in the presence of acids. Acid lability is enhanced by the incorporation of methyl groups, probably because of the

141

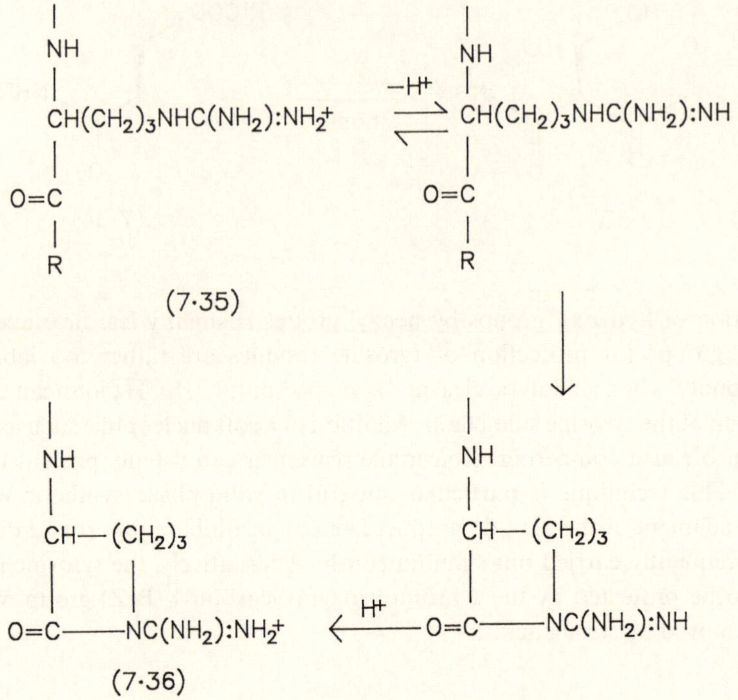

$$NH$$
$$|$$
$$CH(CH_2)_3NHC(NH_2){:}NH_2^+ \quad \underset{\longleftarrow}{\overset{-H^+}{\rightleftharpoons}} \quad CH(CH_2)_3NHC(NH_2){:}NH$$
$$|$$
$$O{=}C$$
$$|$$
$$R$$

(7·35)

$$NH$$
$$|$$
$$CH{-}(CH_2)_3$$
$$|$$
$$O{=}C{-}{-}NC(NH_2){:}NH_2^+ \quad \overset{H^+}{\longleftarrow} \quad O{=}C{-}{-}NC(NH_2){:}NH$$

(7·36)

Scheme 7.7.

greater interaction between aromatic π electrons and vacant d orbitals of sulphur. Thus, the 2,4,6-trimethylbenzenesulphonyl (Mts) (7.37; R¹=H, R²=Me) is introduced using a cheap reagent and is moderately labile with acid. The 4-methoxy-2,3,6-trimethylbenzenesulphonyl group (Mtr) (7.37; R¹=Me, R²=MeO) is much more labile with acid and is widely used. The 4-methoxy-2,3,5,6-tetramethyl-benzenesulphonyl group, surprisingly at first sight, is rather stable. This has been attributed to the presence of the 3,5-methyl groups preventing the MeO group from aligning itself for optimal conjugation with the ring. In order to overcome this, the 2,2,5,7,8-pentamethylchroman-6-sulphonyl group (Pmc) (7.38) was designed, in which the oxygen *para* to the sulphonyl group is locked into the optimal orientation. In consequence, the Pmc group can be quite rapidly removed by 50% $CF_3CO_2H{-}CH_2Cl_2$.

7.5.5 Protection of the imidazole ring of histidine

Protection of the imidazole ring is necessary for several reasons. First, the unprotected heterocycle is sufficiently basic to cause enantiomerisation. Secondly, attempted activation of an α-N-protected histidine derivative can cause lactamisation just like with arginine derivatives (7.39→7.40). Finally, if a carbodiimide

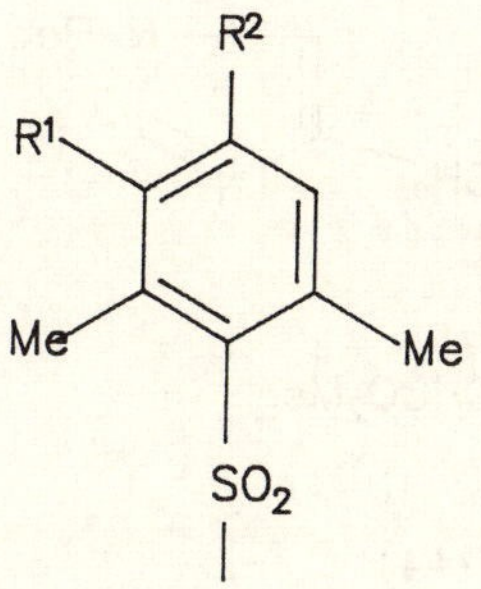

R²
R¹
Me   Me
SO₂
|

(7·37)

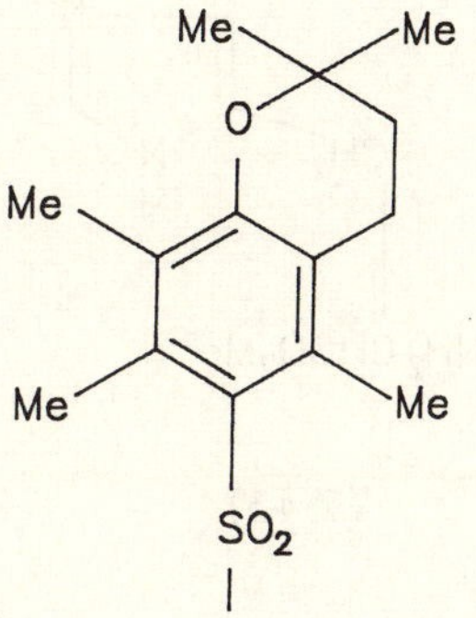

Me   Me
O
Me
Me   Me
SO₂
|

(7·38)

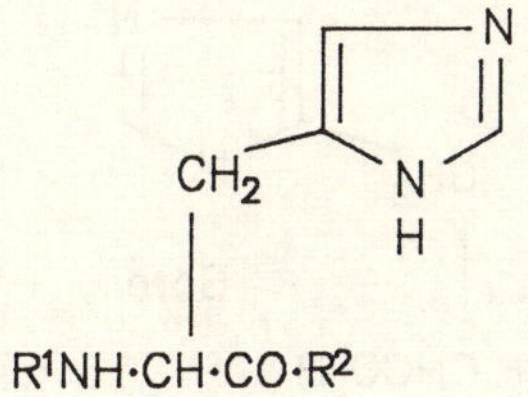

N
CH₂   N
H
|
R¹NH·CH·CO·R²

(7·39)

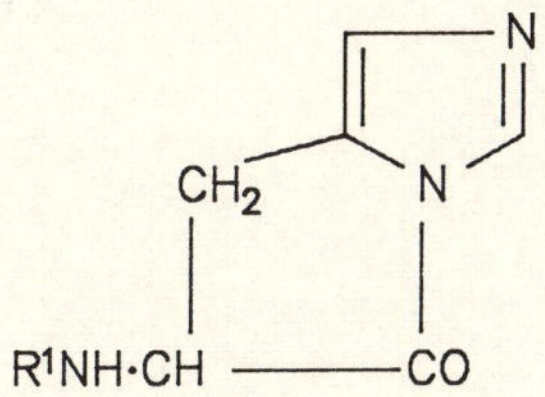

N
CH₂   N
|
R¹NH·CH ——— CO

(7·40)

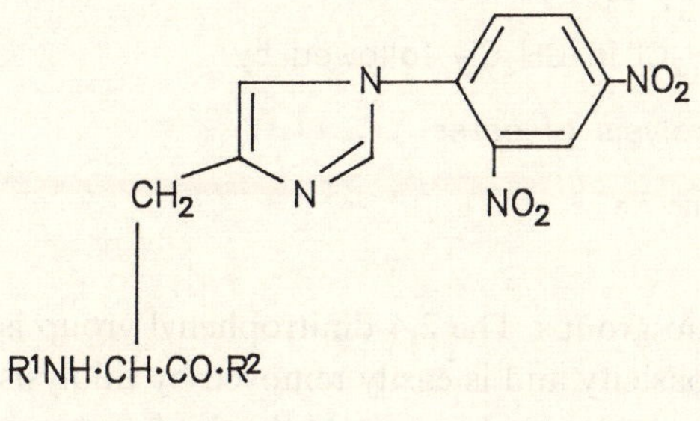

N
CH₂   N
NO₂
NO₂
|
R¹NH·CH·CO·R²

(7·42)

N·C₆H₁₁
N–C
NH·C₆H₁₁
CH₂   N
|
R¹NH·CH·CO·R²

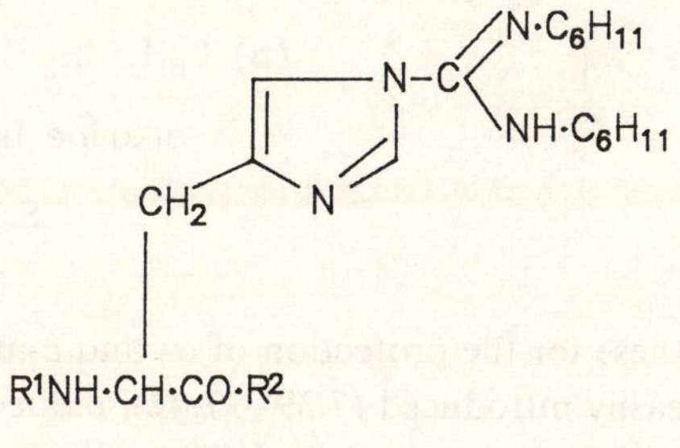

(7·41)

(Section 7.8) is used to activate such a histidine derivative, the latter may undergo amidination (**7.39 → 7.41**).

An early method of protecting the imidazole ring by *N*-benzylation sterically hinders side reactions, but does not decrease the basicity of the ring, so that enantiomerisation remains a serious risk (Section 7.7). An electron-withdrawing group is required, but simple *N*-acyl derivatives are labile and can effect acylation of other groups in the peptide. A urethane group is satisfactory, but it is usual to reserve

143

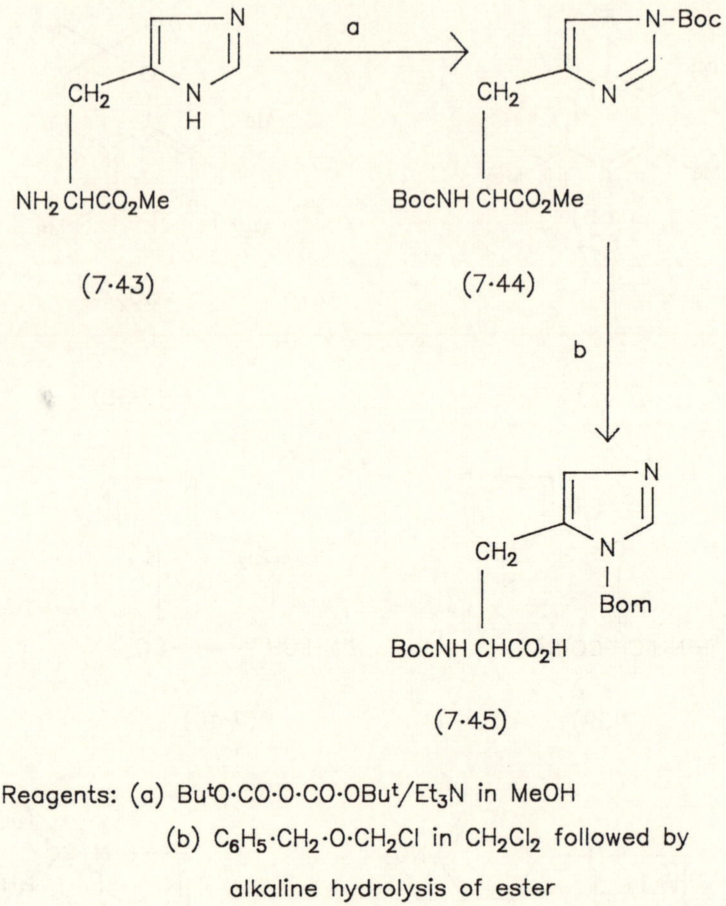
N
a
N–Boc
CH_2
N
H
CH_2
N
NH_2CHCO_2Me
$BocNH\ CHCO_2Me$
(7·43)
(7·44)
b
N
CH_2
N
Bom
$BocNH\ CHCO_2H$
(7·45)
Reagents: (a) $Bu^tO{\cdot}CO{\cdot}O{\cdot}CO{\cdot}OBu^t/Et_3N$ in MeOH
(b) $C_6H_5{\cdot}CH_2{\cdot}O{\cdot}CH_2Cl$ in CH_2Cl_2 followed by
alkaline hydrolysis of ester

Scheme 7.8.

these for the protection of α- and ε-amino groups. The 2,4-dinitrophenyl group is easily introduced (**7.39→7.42**), reduces basicity and is easily removed by thiolysis with 2-mercaptoethanol. The same group can be used to protect the thiol group of cysteine and the hydroxy group of tyrosine, but possesses no special advantages over the more usual blocking groups.

Although a protecting group may enter at either the τ or the π nitrogen atoms (**7.39**), most blocking groups react at the former, but this is less effective than is reaction at the latter in order to diminish enantiomerisation. Fortunately, it has proved possible to block the π nitrogen atom indirectly, either with the benzyloxymethyl (Bom) group (Scheme 7.8) or or with the t-butyloxymethyl (Bum) group (**7.43→7.44 →7.45**). The Bom group is stable against nucleophiles and CF_3CO_2H but is easily cleaved by HBr in CH_3CO_2H and by hydrogenolysis. The Bum group is stable to hydrogenolysis but is cleaved under mildly acidic conditions.

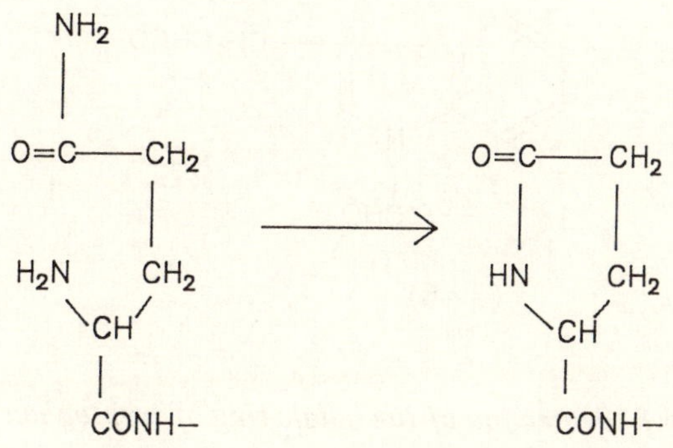

NH$_2$
|
O=C———CH$_2$
|
H$_2$N   CH$_2$
  \CH/
   |
  CONH—

(7·46)

O=C———CH$_2$
| |
HN   CH$_2$
  \CH/
   |
  CONH—

(7·47)

7.5.6 Protection of amide groups

The amide groups of asparagine and glutamine are generally left unprotected unless experiment demonstrates the occurrence of side reactions. Such side reactions usually occur under conditions that favour enantiomerisation, so prevention is considered to be the best policy. When α-N-protected glutamine is coupled to the amino group of a protected amino acid or peptide using N,N'-dicyclohexylcarbodiimide (Section 7.8), conversion of some of the amide into a nitrile can occur. This side reaction can usually be prevented by carrying out the coupling in the presence of N-1-hydroxybenzotriazole (Section 7.8). If a peptide containing a totally unprotected N-terminal glutaminyl residue is left under conditions favouring the unprotected form of the amino group, cyclisation to a 'pyroglutaminyl' peptide may occur (**7.46→7.47**). The amide groups both of asparagine and of glutamine can be protected satisfactorily by the trityl (Trt) group. The Trt group is introduced using TrtOH–(CH$_3$CO$_2$)$_2$O–CH$_3$CO$_2$H–H$_2$SO$_4$ at 50 °C. It is stable to base and to hydrogenolysis and is removed by CF$_3$CO$_2$H.

7.5.7 Protection of the thioether side-chain of methionine

Protection of the thioether group of methionine is desirable for two reasons. First, oxidation to sulphoxide occurs slowly during repeated manipulation of solutions exposed to air. Secondly, acidolytic release of carbocations during peptide synthesis readily leads to the formation of sulphonium derivatives. It is customary to start with the sulphoxide prepared, for example, by mild oxidation with hydrogen peroxide to pre-empt the first difficulty and to prevent the second. Although diastereoisomeric sulphoxides are probably formed, this does not lead to any complications. The sulphoxide can be reduced back to the thioether at the end of the synthesis by a thiol such as 2-mercaptoethanol, although more esoteric reagents such as selenophenol and 2-mercaptopyridine act more quickly.

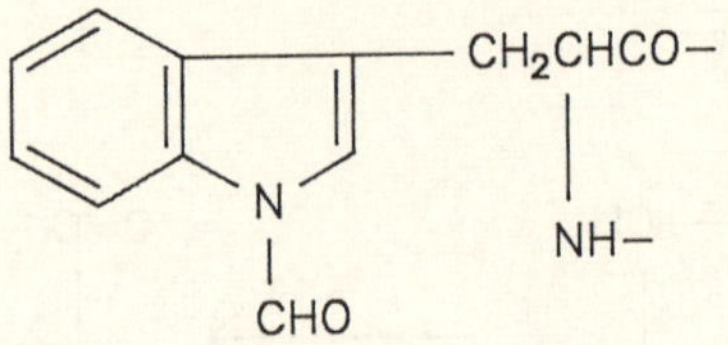

(7·48)

7.5.8 Protection of the indole ring of tryptophan

The indole ring of tryptophan is rather unstable to acids. It is prone to oxidation and to *N*-substitution by carbocations. Protection can be afforded by a formyl group on the nitrogen atom of the indole ring (**7.48**). The group is introduced with $HCO_2H–HCl$ and removed at pH 9–10.

7.6 Deprotection procedures

Much of peptide synthesis is concerned with the use and removal of protecting groups. The α-*N*-protecting group on the current *N*-terminal residue must be removed before the next amino acid can be coupled and various methods for doing this have been described (Section 7.3). At the end of the synthesis, all the protecting groups must be removed and, if solid-phase methodology (Section 7.9) has been used, the bond between the *C*-terminal residue and the insoluble matrix must be cleaved. Exposure to a strong acid (e.g. HF) is commonly used. Unfortunately, this favours an S_N1 mechanism and the formation of carbocations (Section 7.5) that can alkylate the product at susceptible points. It is therefore desirable to use conditions that favour an S_N2 mechanism. This can be achieved by using a combination of a strong (hard) acid and a soft nucleophile so that a push–pull mechanism (e.g. **7.49** + **7.50** → **7.51**) can operate. The soft nucleophile is a scavenger for carbocations. If necessary, more concentrated HF can be used afterwards to remove any surviving protecting groups. Potent trimethylsilylating reagents such as $CF_3SO_3SiMe_3$ can replace the proton and a mixture of this and CF_3CO_2H with PhSMe as the soft nucleophile removes protecting groups rapidly, giving a clean product.

7.7 Enantiomerisation[2] during peptide synthesis

The proton on C_2 of an amino-acid derivative is slightly labilised by the adjacent carbonyl group, especially under alkaline conditions. This causes slow enantiomerisation (**7.52** → **7.53** → **7.54**) (Scheme 7.9).

[2] Most texts describing peptide synthesis use the term 'racemisation' when referring to loss of chirality at one chiral centre when two or more chiral centres are present. Benoiton (1994) has pointed out that this practice is incorrect and that 'enantiomerisation' should be used.

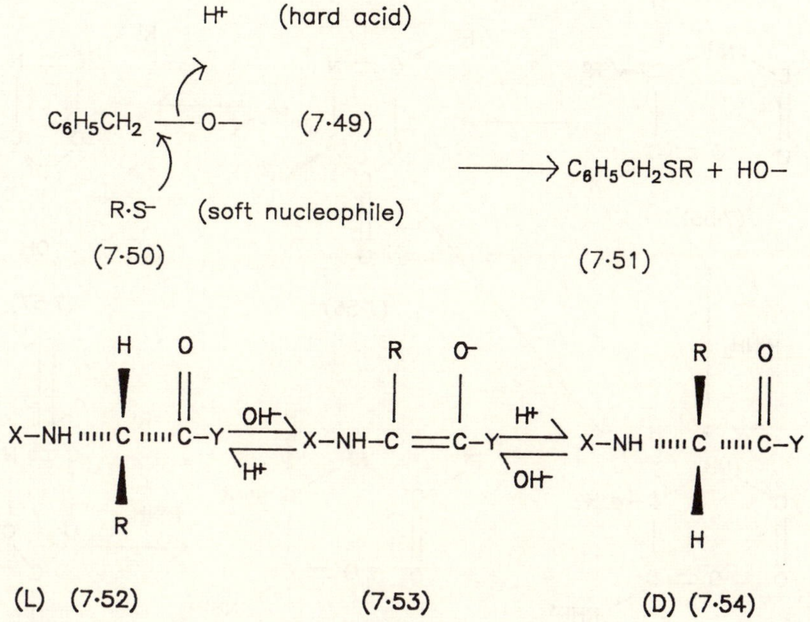

H⁺ (hard acid)

$$C_6H_5CH_2 \longrightarrow O— \quad (7·49)$$

$$\longrightarrow C_6H_5CH_2SR + HO—$$

R·S⁻ (soft nucleophile)

(7·50)

(7·51)

Scheme 7.9.

(L) (7·52)   (7·53)   (D) (7·54)

A more important cause of enantiomerisation stems (Scheme 7.10) from the cyclisation of carboxy-activated derivatives of α-*N*-acylamino acids including peptides (7.55) to form a 5(4)-oxazolone (7.56) concurrently with the formation of a coupled product (7.55→7.60) Enolisation of the 5(4H)-oxazolone (7.56⇌ 7.57⇌7.58) destroys the chirality and the rate of enantiomerisation depends on the relative rates of the steps (7.55→7.56⇌7.57⇌7.58). The other mechanism of enantiomerisation may operate concurrently. Enantiomerisation is favoured if R¹CO is small and electron-attracting (e.g. CH_3CO). The presence of the Boc group (R¹=Me₃CO) militates against enantiomerisation on both counts.

Clearly, formation of the 5(4)-oxazolone will be favoured if X is a good leaving group, but this is also one factor that favours peptide-bond formation. Coupling and enantiomerisation are also favoured by the use of polar solvents, since this favours the departure of X. Thus, every new protecting group or method of coupling must be validated by subjecting it to quantitative stereochemical testing under adverse conditions. One such test involves coupling Z—Gly—Phe—OH to H— Gly—OEt. The racemic form of Z—Gly—Phe—Gly—OEt is easily separated from the L enantiomer by fractional crystallisation from ethanol. Less than 1% of the racemate can be detected by this method. There is a related test in which Bz—L— Leu—OH is coupled to Gly—OEt.

Enantiomerisation can be detected by using an analytical procedure such as HPLC with high resolution and sensitivity to separate and quantify diastereoiso-

147

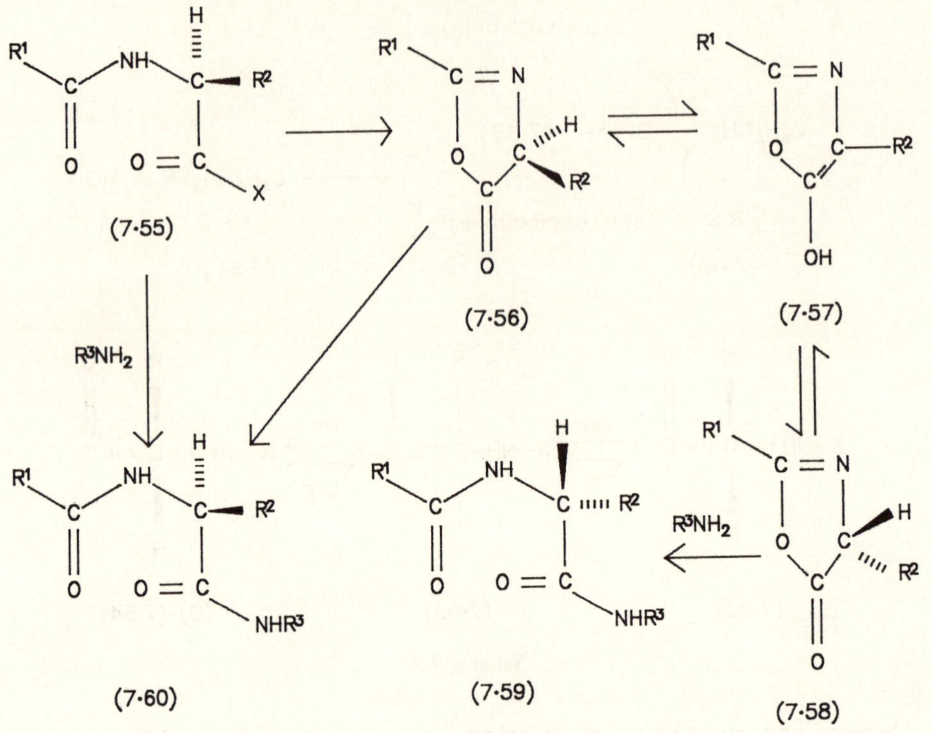

Scheme 7.10.

meric derivatives. For example, if Z—Gly—L—Ala—OH is coupled to H—L—Leu—OBzl and protecting groups are removed by hydrogenolysis, the product, H—Gly—L—Ala—L—Leu—OBzl, and the diastereoisomer containing D-Ala, formed as a result of enantiomerisation during coupling, can be separated easily by chromatography on the cation-exchange resin Dowex 50 (Izumiya and Muraoka, 1969). Again, the diastereoisomers formed in the synthesis of H—Phe—Phe—Ala—OBzl can be separated by HPLC.

An elegant method for studying the extent of enantiomerisation during coupling makes use of the absolute stereospecificity of a proteolytic enzyme such as leucine aminopeptidase (Bosshard *et al.*, 1973). Peptides in which the residue adjacent to the N-terminus has the D configuration are totally resistant to this enzyme. On the other hand, peptides with a hydrophobic L-amino acid at the N-terminus and an L-amino acid other than Pro at the adjacent position are hydrolysed at the first peptide bond. Thus, if Z—L—Ala—D—Ala—OH is coupled to H—L—Ala—L—Ala—OBzl and protecting groups are removed by hydrogenolysis, the product should be totally resistant to leucine aminopeptidase. If enantiomerisation occurs, however, the all-L peptide is completely digested and 4 moles of Ala are produced per mole of racemic product, the amplification of the signal increasing the sensitivity of the test.

The factors that determine the extent of enantiomerisation (Kemp, 1979) include (a) the structure of the group attached to the α-amino group of the next residue to be coupled, (b) the structure of the next residue to be coupled, (c) the coupling procedure, (d) the choice of solvent and (e) control of the temperature. The first of these factors is discussed here; the remainder are considered in the section on peptide-bond formation (Section 7.8). Obviously, a powerful electron-withdrawing group attached to the α-amino group will labilise the proton on the α-carbon atom. For this reason, the *N*-trifluoroacetyl group is never used to protect the α-amino group, although it is often useful to protect the ε-amino group of lysine. As mentioned above, the more important route for enantiomerisation involves the formation of a 5(4H)-oxazolone. This process will be favoured by (a) a powerful electron-withdrawing group for activating the carboxy group in order to form the next peptide bond and (b) a substantial electron density on the carbonyl oxygen atom of the group used to protect the α-amino group. These two factors favour the cyclisation step (**7.55→7.56**). The use of a urethane group to protect the α-amino group disperses the electron density between two oxygen atoms. Consequently, Z, Boc, Fmoc and Alloc groups favour the retention of optical purity whereas simple acyl groups, including the peptide bond itself, permit enantiomerisation via the oxazolone route, especially under basic conditions.

At this point, it should be obvious why it is usual to couple one amino acid at a time, each with its amino group protected by a urethane moiety, building from the *C*-terminus to the *N*-terminus. It is a revealing commentary on the stereospecificity and control of enzyme-mediated processes to realise that the biosynthesis of a protein proceeds in the opposite direction. Obviously, peptides can be coupled chemically without risk of enantiomerisation if the *C*-terminal residue of the intended *N*-terminal fragment is Gly since this is not chiral. *C*-Terminal Pro also usually offers considerable protection against enantiomerisation in fragment coupling. Can you suggest why this might be so?

7.8 Methods for forming peptide bonds

Although published methods for extending a peptide chain are legion (Jones, 1979, 1994), most are unused except perhaps by their inventors. At first sight, it may seem surprising that so much research effort has been directed towards the synthesis of a particular type of amide and that so few methods measure up to requirements. There are three main considerations, (i) minimal enantiomerisation, (ii) high yield and (iii) rapid coupling. The first of these has been considered in general terms (Section 7.7); the second was included under basic strategy (Section 7.1) and is essential for solid-phase peptide synthesis (Section 7.9). The third consideration is important after (i) and (ii) have been secured.

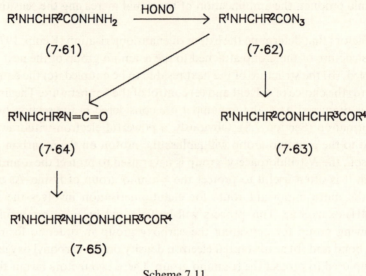

Scheme 7.11.

7.8.1 *The acyl azide method*

This is by far the oldest method of forming peptide bonds that is still in use (Scheme 7.11) and it depends on the production of an acyl azide from an acyl hydrazide by the classical Curtius procedure (**7.61**→**7.62**→**7.63**) (Meienhofer, 1979a). The generation of an acyl azide from an acyl hydrazide used to be effected with aqueous sodium nitrite and a mixture of aqueous acetic and hydrochloric acids but nowadays a non-aqueous system is preferred and amyl nitrite replaces sodium nitrite. Several side reactions are possible, including the rearrangement of the acyl azide (**7.62**) to the isocyanate (**7.64**). The latter can react additively with the amino group of an amino-acid ester to give a urea (**7.65**). Separation of this from the desired peptide is often very difficult on the preparative scale. Coupling reactions of acyl azides are quite slow and this may account for the one redeeming feature of the Curtius procedure, namely that there is usually a high degree of retention of optical purity. A highly reactive derivative of a carboxylic acid such as an anhydride, aryl ester or *O*-acylisourea is liable to form the corresponding 5(4H)-oxazolone with an increased risk of enantiomerisation (Section 7.7). Acyl azides are much less reactive and would probably be even less so were it not for a possible intramolecular base catalysis (**7.66**→**7.67**→**7.68**) that would be expected to prevent formation of the 5(4H)-oxazolone. Despite this advantage, the possible side reactions and slowness of coupling on the one hand and the improvement of competing faster methods on the other have more or less rendered the Curtius method obsolete, although it is still frequently used for coupling protected peptides to yield a bigger molecule (fragment coupling).

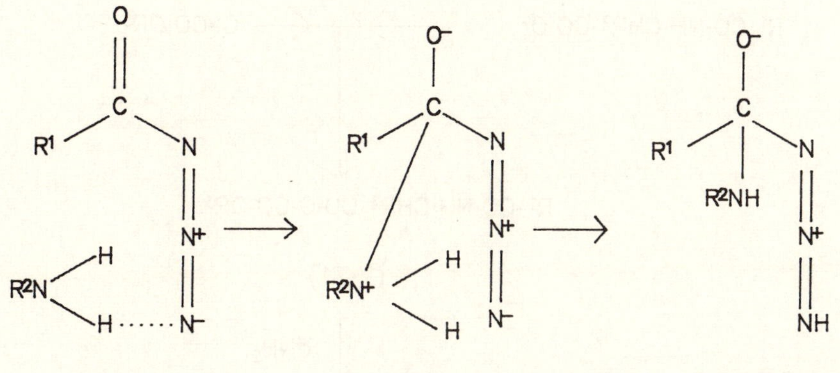

(7·66) (7·67) (7·68)

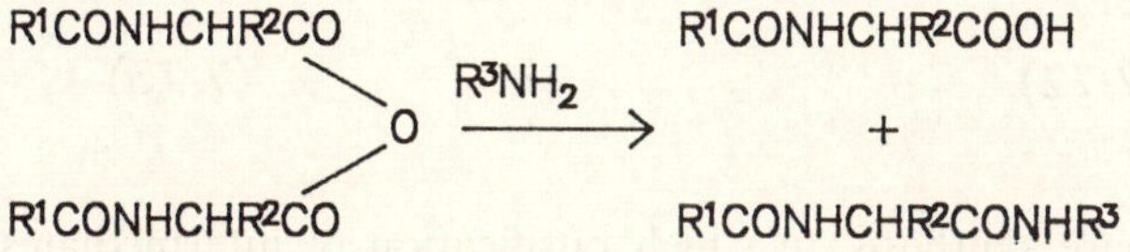

R¹CONHCHR²CO

O $\xrightarrow{R^3NH_2}$

R¹CONHCHR²CO

(7·69)

R¹CONHCHR²COOH

+

R¹CONHCHR²CONHR³

(7·70)

7.8.2 The use of acid chlorides and acid fluorides

The prospect of using acid chlorides and fluorides as intermediates for peptide-bond formation would have been almost laughable until recently because of the risk of enantiomerisation (Section 7.7). Boc and Fmoc amino-acid fluorides, however, are rather stable and can be prepared by the interaction of cyanuric fluoride and the corresponding acid (Carpino *et al.*, 1990a). Fmoc amino-acid chlorides, although they are less stable and react more slowly than do the fluorides, can be obtained from the reaction between unsymmetrical acid anhydrides (see below) and dry HCl (Chen *et al.*, 1991). The preferred base for use in these reactions is tris(2-aminoethyl)amine (Carpino *et al.*, 1990b). *N,N*-Bis-Boc-amino acids (Section 3.1) give acyl fluorides when treated with cyanuric fluoride in CH_2Cl_2 at $-30\,°C$ (Savrda and Wakselman, 1992; Carpino *et al.*, 1993). Although the presence of two bulky Boc groups on the nitrogen atom might have been anticipated to cause unacceptable steric hindrance, these derivatives are very suitable for peptide synthesis.

7.8.3 The use of acid anhydrides

The use of symmetrical anhydrides of α-*N*-protected amino acids (**7.69**) was originally considered unattractive because only half the derivative was converted into peptide (**7.69→7.70**). On the other hand, symmetrical acid anhydrides, unlike unsymmetrical acid anhydrides, give only a single product and this is particularly

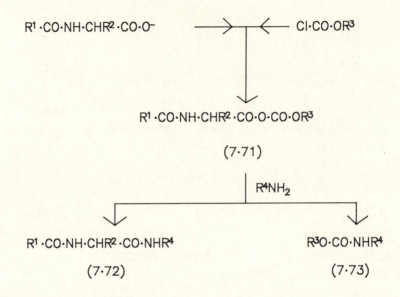

$$R^1 \cdot CO \cdot NH \cdot CHR^2 \cdot CO \cdot O^- \qquad \longrightarrow \qquad \longleftarrow \qquad Cl \cdot CO \cdot OR^3$$

$$R^1 \cdot CO \cdot NH \cdot CHR^2 \cdot CO \cdot O \cdot CO \cdot OR^3$$

$$(7 \cdot 71)$$

$$R^4NH_2$$

$$R^1 \cdot CO \cdot NH \cdot CHR^2 \cdot CO \cdot NHR^4 \qquad\qquad R^3O \cdot CO \cdot NHR^4$$

$$(7 \cdot 72) \qquad\qquad\qquad\qquad (7 \cdot 73)$$

important in solid-phase synthesis, in which purification of intermediates is not carried out.

In contrast, unsymmetrical anhydrides[3] have been used for many years (Meienhofer, 1979b). The expectation that the anhydride would be cleaved so that the leaving group is derived from the stronger of the two acids forming the anhydride is not always borne out. Steric factors, solvent polarity and the relative thermo-dynamic stability of the two acids can be of crucial importance. Moreover, there is always a risk that unsymmetrical acid anhydrides can disproportionate to two sym-metrical anhydrides with consequent lowering of the yield of the desired product.

Unsymmetrical acid anhydrides are obtained by reaction at low temperature of N-protected amino acids and acyl halides, especially chloroformate esters, in the presence of a tertiary base. The anhydride (**7.71**) is then allowed to react with an amino acid or with a peptide or ester thereof. The choice of a chloroformate with a bulky alkyl group (e.g. isobutyl or *sec*-butyl) helps to direct the reaction to give the desired product (**7.72**) rather than the alternative urethane derivative (**7.73**). Cleavage of the anhydride in the desired mode and with minimal enantiomerisation is favoured by using a solvent of low polarity. Retention of chiral purity is also favoured by using reaction times as short as possible consistent with obtaining good yields. As an example of an unsymmetrical anhydride that is not derived from a car-boxylic acid, diphenylphosphinic chloride (**7.74**) gives anhydrides that cleave in the desired direction without disproportionation (Ramage *et al.*, 1985).

---

[3] Most peptide chemists refer to unsymmetrical anhydrides as mixed anhydrides. This title is best reserved for those cases in which unsymmetrical anhydrides undergo disproportionation to give a mixture of anhydrides:

$$2R^1COOCOR^2 \rightleftharpoons R^1COOCOR^1 + R^2COOCOR^2.$$

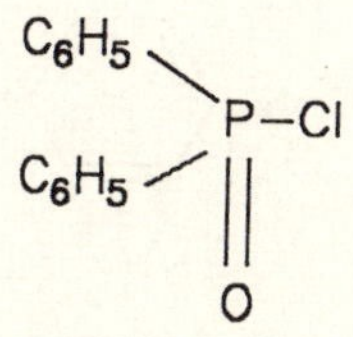

$$C_6H_5 \diagdown \atop C_6H_5 \diagup P-Cl \atop \overset{\|}{O}$$

(7·74)

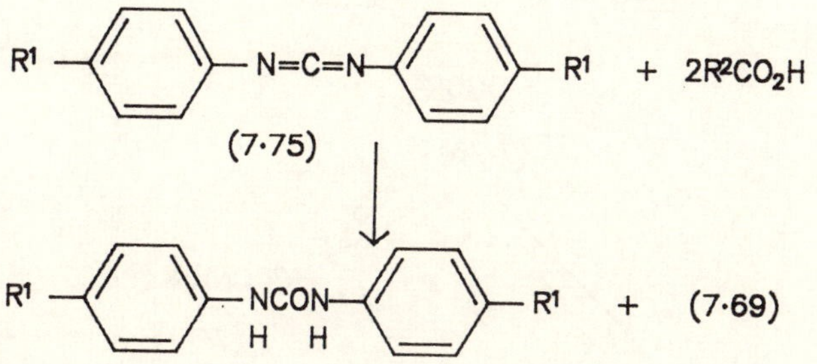

$$R^1 - \hspace{-0.5em}\bigcirc\hspace{-0.5em} - N=C=N - \hspace{-0.5em}\bigcirc\hspace{-0.5em} - R^1 \quad + \quad 2R^2CO_2H$$

(7·75)

$$R^1 - \hspace{-0.5em}\bigcirc\hspace{-0.5em} - \underset{H \; H}{NCON} - \hspace{-0.5em}\bigcirc\hspace{-0.5em} - R^1 \quad + \quad (7·69)$$

7.8.4 The use of carbodiimides

Depending on their structure and on conditions of reaction, carbodiimides (7.75) react with carboxylic acids to give symmetrical anhydrides (7.69) or O-acylisoureas (7.76) (Rich and Singh, 1979) (Scheme 7.12). The latter react with esters of amino acids or peptides to give peptide derivatives directly or with phenols to give aryl esters (7.76→7.77). Although this route avoids the possibility of wrong cleavage found with unsymmetrical anhydrides, there is a separate complexity. O-Acylisoureas tend to rearrange to N-acylureas (7.78). The latter are unreactive towards amines and may be difficult to separate from products. Enantiomerisation is a further hazard, but maintenance of a low temperature and use of non-polar solvents for couplings, if this is feasible, usually gives satisfactory results.

7.8.5 The use of reactive esters

Aryl and other reactive ('activated') esters can be prepared from O-acyl isourea intermediates as outlined above (Bodanszky, 1979). The esters can often be isolated and stored for subsequent use or can be generated and used in situ in a coupling reaction. The reactivity of the ester is related to the pK_a of the corresponding phenol; the reactivity generally increases as the pK_a decreases. Indeed, aryl esters can be regarded as unsymmetrical acid anhydrides that can be cleaved in one direction only. One potential complication is thus avoided, although the dangers of enantiomerisation and formation of N-acylureas remain. p-Nitrophenol, 2,4,5-trichlorophenol, pentachlorophenol and pentafluorophenol have all been used to generate reactive esters for peptide synthesis. Several N-hydroxy heterocyclic compounds are also particularly useful. Thus, N-hydroxysuccinimide, which is a hydroxamic acid (pK_a 4.3),

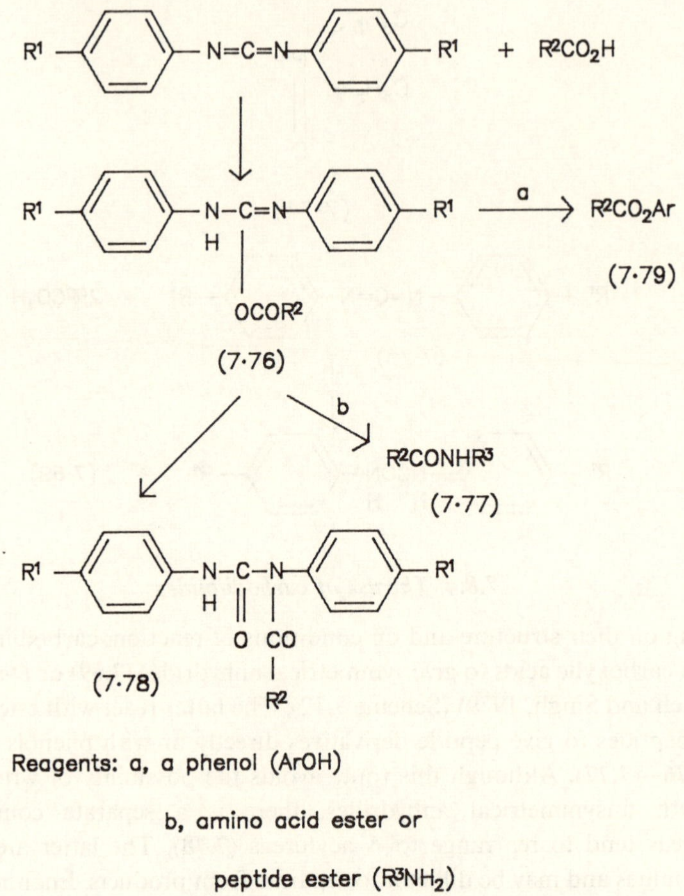

$$R^1 \text{—} \bigcirc \text{—} N{=}C{=}N \text{—} \bigcirc \text{—} R^1 \quad + \quad R^2CO_2H$$

$$R^1 \text{—} \bigcirc \text{—} \underset{H}{N}\text{—}C{=}N \text{—} \bigcirc \text{—} R^1 \quad \xrightarrow{a} \quad R^2CO_2Ar$$

$$\underset{OCOR^2}{\big|}$$

(7·79)

(7·76)

$$R^2CONHR^3$$

b

(7·77)

$$R^1 \text{—} \bigcirc \text{—} \underset{H}{N}\text{—}\underset{\underset{O}{\|}}{C}\text{—}\underset{\underset{CO}{|}}{N} \text{—} \bigcirc \text{—} R^1$$

(7·78)

$$R^2$$

Reagents: a, a phenol (ArOH)

b, amino acid ester or

peptide ester (R³NH₂)

Scheme 7.12.

gives rise to reactive esters when it is acylated by N-protected amino acids in the presence of carbodiimides. The reactivity of such esters is not simply due to their anhydride character. Reaction of the esters with amino compounds is subject to anchimeric base catalysis (7.79→7.80). The importance of anchimeric assistance can be appreciated since esters of N-hydroxypiperidine, which can hardly be regarded as acid anhydrides, also undergo ready aminolysis. In the case of N-hydroxysuccinimide, the esters are sufficiently stable to store and, when they are used to couple with an amino compound, the liberated N-hydroxysuccinimide is water-soluble and can be easily removed by washing.

The existence of anchimeric assistance during coupling reactions involving esters of N-hydroxy-heterocycles has proved to be extremely useful in peptide synthesis. Whether one is starting with an N-protected amino acid and a carbodiimide or using a reactive ester, either previously isolated or generated in situ, the addition of

154

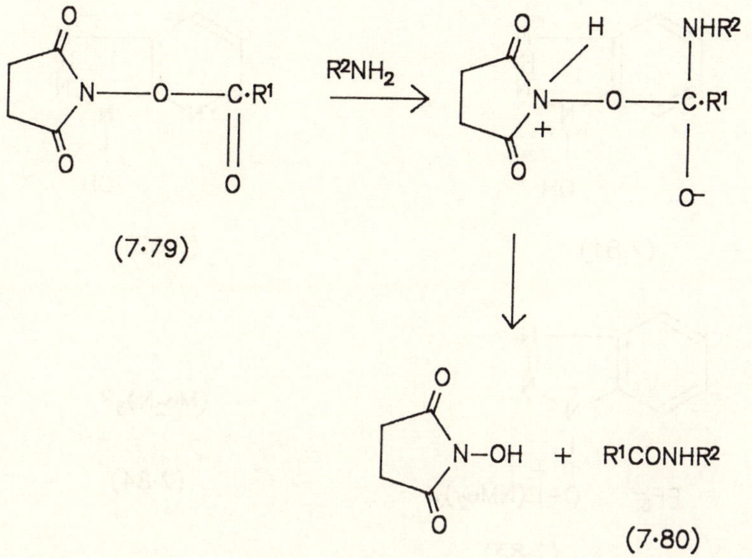

$$(7 \cdot 79)$$

$$(7 \cdot 80)$$

a suitable *N*-hydroxy-heterocycle can accelerate coupling and decrease enantiomerisation. 1-*N*-hydroxybenzotriazole (**7.81**) is a favourite additive and 1-*N*-hydroxy-7-azabenzotriazole (**7.82**) is a more recent and more effective additive (Carpino, 1993). The mechanism of these reactions has not been identified definitively. It is possible that esters of the *N*-hydroxy compounds may be formed by coupling of an *N*-protected amino acid in the presence of a carbodiimide or by anchimerically assisted *trans*-esterification of another reactive ester.

7.8.6 *The use of phosphonium and isouronium derivatives*

A particularly useful coupling agent, benzotriazol-1-yl trisdimethylaminophosphonium hexafluorophosphate (BOP) (**7.83**) can easily be prepared by the interaction of 1-*N*-hydroxybenzotriazole (**7.81**), trisdimethylaminophosphine (**7.84**) and carbon tetrachloride at low temperature followed by treatment with potassium hexafluorophosphate. Reaction of BOP with the anion of a protected amino acid probably yields the corresponding ester. BOP has been found to be effective in difficult couplings involving, for example, amino acids with bulky side-chains and it causes very little enantiomerisation. It is also very effective in solid-phase peptide synthesis (Section 7.9). Unfortunately, a possible starting material for the preparation of BOP and a product from a peptide synthesis involving its use is $(Me_2N)_3PO$, which is suspected to be a carcinogen. The closely related reagent (**7.85**) (Coste *et al.*, 1990) is a satisfactory replacement for BOP since the corresponding starting reagent and product after coupling, tris(pyrrolidino) phosphine oxide, is believed not to be carcinogenic.

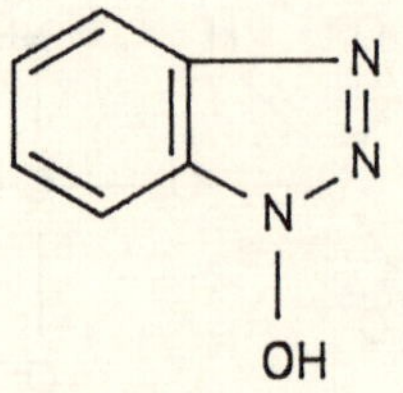

(7·81)

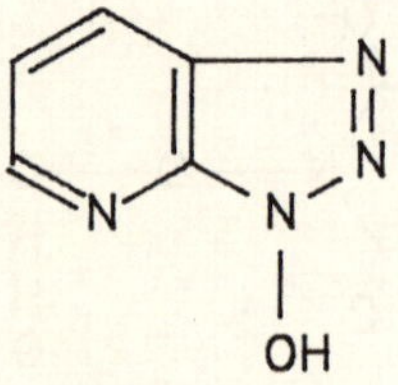

(7·82)

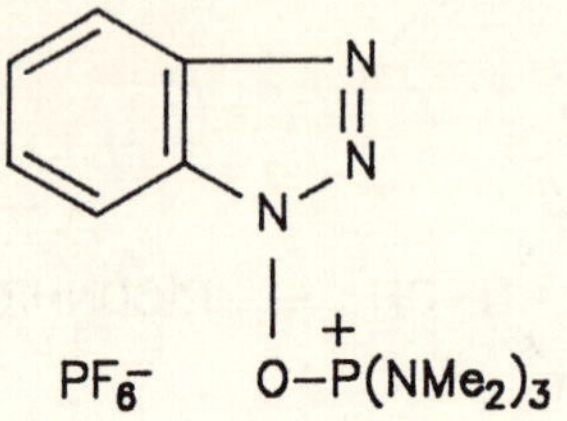

PF_6^- $O-\overset{+}{P}(NMe_2)_3$

(7·83)

$(Me_2N)_3P$

(7·84)

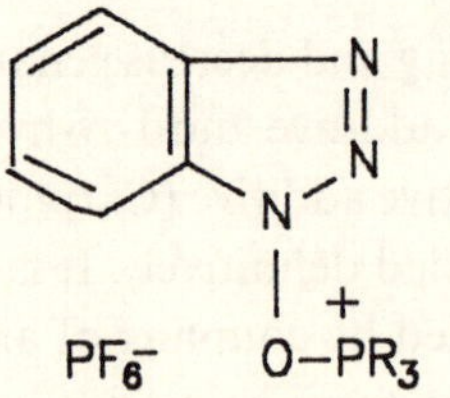

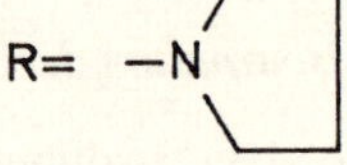

PF_6^- $O-\overset{+}{P}R_3$

PF_6^- $O-\overset{+}{C}R_2$

R=—NMe$_2$, —N◁

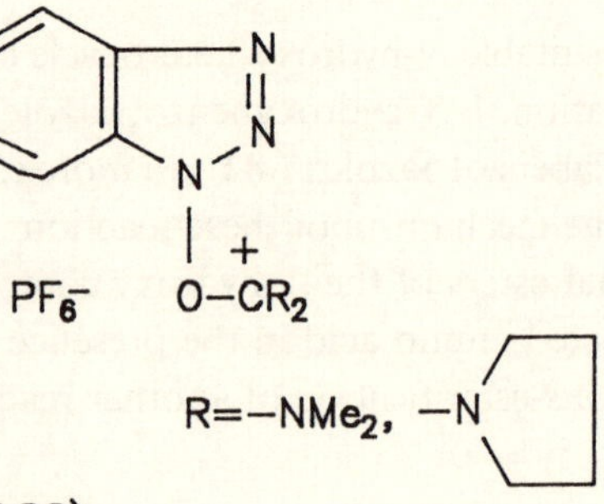

(7·86)

(7·85)   R= —N◁

Formally related reagents for peptide synthesis are uronium derivatives such as (**7.86**) (Knorr *et al.*, 1989; Chen and Xu, 1992). Like the phosphonium derivatives, this type of reagent causes very little enantiomerisation.

7.9 Solid-phase peptide synthesis (SPPS)

There are two steps, deprotection and coupling, involved in the stepwise addition of amino acids to a growing peptide chain. When using conventional methods with equimolar amounts of reagents, it is necessary to purify the product at each stage by techniques such as washing, crystallisation and chromatography. The synthesis of peptides containing quite a small number of amino acids can consequently be laborious with low overall yields. In earlier times, such work justifiably evoked considerable acclaim. For example, the synthesis of the nonapeptides, oxytocin, vasopressin and some analogues, earned du Vigneaud a Nobel prize in 1955. The advent of SPPS, which earned another Nobel prize (Merrifield, 1986), and its subsequent technical improvements (Stewart and Young, 1984; Atherton and Sheppard, 1989) have brought such syntheses within the scope of an undergraduate chemist or biochemist. The reasons are simple to understand in principle. The synthesis is carried out on an insoluble solid matrix that is freed from soluble by-products and excess reagents by washing with suitable solvents. Protecting groups on the side-chains of amino acids are retained until the end of the synthesis and the peptide is finally detached from the insoluble support and purified by a suitable procedure such as HPLC or ion-exchange chromatography. Several commercial instruments are available for carrying out most of the steps under computer control. It is essential, however, for yields to approach 100% as closely as possible at every step. Reference to the beginning of this chapter (Section 7.1) will show that a yield of 80% in the addition of each amino acid gives an overall yield of 21% of an octapeptide. In other words, 79% of the product consists mainly of incomplete sequences including eight possible heptapeptides, twenty-eight possible hexapeptides and so on. With such an intractable mixture, all the potential advantages of the solid-phase methodology would be lost. In spite of the extra cost, therefore, it is imperative to use a substantial excess of the N-protected amino acid or, frequently, double coupling sequences in order to achieve a near quantitative yield at each step. It is possible to test for completion of the acylation step by subjecting a small sample of the resin to the ninhydrin test for amino compounds. A negative test result indicates that acylation has proceeded to completion. Application of this test involves interruption of the computer-controlled synthesis. More convenient continuous methods for following the course of the acylation steps during SPPS are described later in this section. As a compensation for the need to use a substantial excess of reagents in SPPS, it is feasible to work on a much smaller scale than with conventional synthetic methods because there are no losses by manipulation until the product is detached from the matrix at the end of the synthesis. It is also possible to recover the excess of reagent used at each acylation step if desired, although few teams would regard this as a cost-effective measure.

Numerous organic polymers have been designed for solid-phase synthesis, but the most popular are based on polystyrene (**7.87**) or polyacrylamide (**7.88**), incorporating an appropriate amount of the crosslinking agent. Remarkable successes have been achieved with both supports. It is essential that the resin should swell in organic solvents to permit access to groups within the pores of the resin by chemical reagents during cycles of deprotection, coupling and washing operations. The degree of swelling should be commensurate with the increase in molecular size as peptide assembly proceeds. The ability of a resin to swell is controlled by the amount of the cross linking agent used in its preparation. In the case of the polyacrylamide type of resin, it has been found advantageous to enclose the polymer in highly porous par-

$$
\begin{array}{cc}
\left[\begin{array}{c} \text{CH} \\ | \\ \text{CH}_2 \\ | \end{array} \!\!-\!\! \bigcirc \right]_n & \left[\begin{array}{c} \text{CHCONH}_2 \\ | \\ \text{CH}_2 \\ | \end{array} \right]_n \\
(7\cdot87) & (7\cdot88)
\end{array}
$$

ticles of inert inorganic materials such as keiselguhr. This prevents the resin from collapsing and blocking the flow of liquid through the particles. It has been fairly standard practice to use the Boc protecting group with the polystyrene matrix and the Fmoc protecting group with the polyacrylamide support, but this is a reflection of the predilections of the main innovators rather than being based on any real chemical restrictions. Typical protocols are given for both methods (Schemes 7.13 and 7.14). Design of the linker molecule and the mode of attachment of the peptide are important. Acidolytic methods are generally favoured for detaching the peptide from the resin at the end of the synthesis and, since the use of the Boc group requires repeated exposure to mildly acidic conditions, the peptide is secured to the support by a bond that requires treatment with strong acids such as hydrofluoric and trifluoromethanesulphonic acids to effect detachment. Use of the Fmoc group, in contrast, requires exposure to a base for deprotection during each cycle and hence the bond securing the peptide to the resin can be of a type that is cleaved by somewhat milder acidic conditions (Atherton et al., 1981).

Much of the success of SPPS has resulted from the ingenious design of linkers between the C-terminal residue of the peptide and the resin. As indicated above, the bond between a peptide and a linker must withstand the cycle of deprotection and coupling and yet be cleaved under conditions as mild as possible at the end of the synthesis so that the peptide is neither degraded nor enantiomerised. In the classical Merrifield method using Boc groups for protecting α-amino groups, the peptide–resin link is repeatedly subjected to acidic conditions during deprotection steps. Benzyl esters are slightly labile under acidic conditions and probably about 1% of the peptide was lost on each exposure during the early development of the Merrifield method. This was an unacceptable loss in the synthesis of long peptides and so the bond between the peptide and the linker required stabilisation. The resistance to acid was increased by a factor of 100 by introducing the electron-withdrawing phenylacetamidomethyl group (Pam) to give the resin (**7.89**). Pam resins have been used successfully for the synthesis of large peptides (Scheme 7.13), including, for example, interleukin 3, which contains 140 amino-acid residues.

The Merrifield school has also designed linkers that permit peptides to be detached by different methods (Tam et al., 1981). A linker based on the benzhydry-

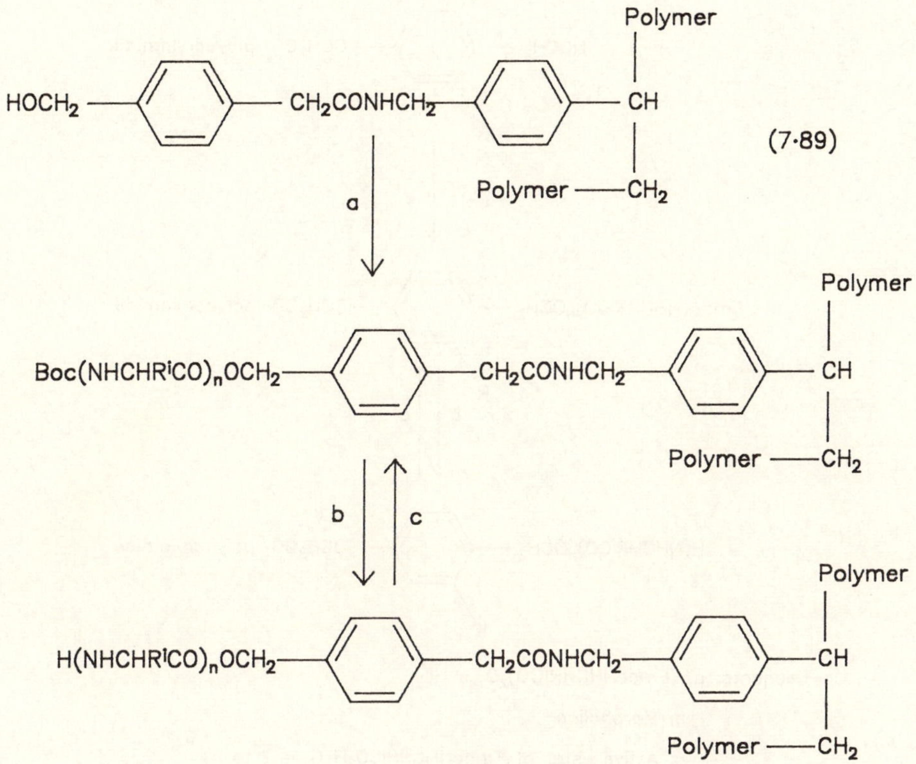

$$\text{HOCH}_2 - \bigcirc - \text{CH}_2\text{CONHCH}_2 - \bigcirc \overset{\overset{\text{Polymer}}{|}}{\underset{\underset{\text{Polymer} - \text{CH}_2}{|}}{\text{CH}}} \quad (7 \cdot 89)$$

a ↓

$$\text{Boc(NHCHR}^i\text{CO)}_n\text{OCH}_2 - \bigcirc - \text{CH}_2\text{CONHCH}_2 - \bigcirc \overset{\overset{\text{Polymer}}{|}}{\underset{\underset{\text{Polymer} - \text{CH}_2}{|}}{\text{CH}}}$$

b ↓ ↑ c

$$\text{H(NHCHR}^i\text{CO)}_n\text{OCH}_2 - \bigcirc - \text{CH}_2\text{CONHCH}_2 - \bigcirc \overset{\overset{\text{Polymer}}{|}}{\underset{\underset{\text{Polymer} - \text{CH}_2}{|}}{\text{CH}}}$$

Reagents: a, $\text{BocNHCHR}^i\text{CO}_2\text{H}$/carbodiimide (i=n=1)

b, $\text{CF}_3\text{CO}_2\text{H}$ followed by teriary base

c, Reactive derivative of $\text{BocNHCHR}^i\text{CO}_2\text{H}$ (i=2 to n)

Scheme 7.13. Solid-phase synthesis with Pam linker on polystyrene resin. Note that polystyrene resin is crosslinked by incorporating a small quantity of divinylbenzene.

lamine structure (7.90) allows the peptide to be detached as the amide in two different ways (Scheme 7.15), one of which does not require a strong acid such as HF. Linkers have also been designed to allow the peptide to be detached by exposure to the fluoride ion (e.g. 7.91) (Ramage *et al.*, 1992).

In many cases, it is advantageous if the conditions used to cleave the peptide from the resin also remove some or all of the protecting groups on the side-chains of the constituent amino acids. This would obviously be undesirable if solid-phase methodology were used for the synthesis of fragments of a large protein destined for fragment condensation by classical methods. It is also important that the linker shall neither impede the approach of reagents used during the synthetic cycle nor cause

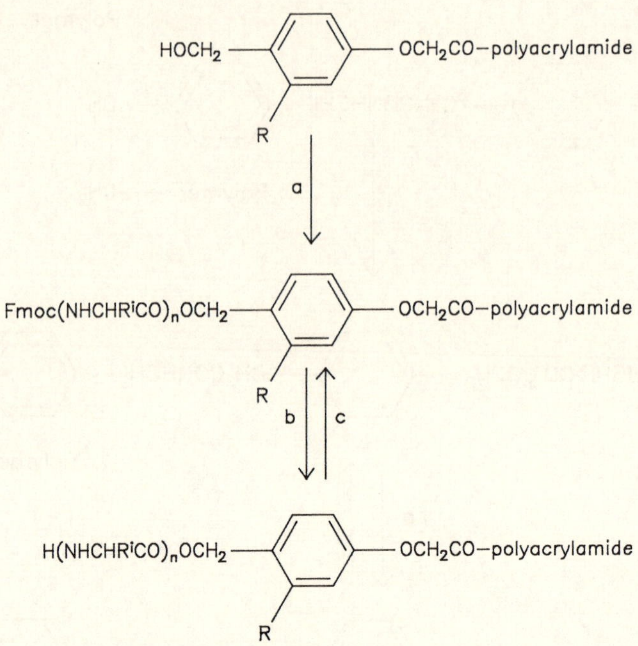

HOCH₂—⟨benzene ring with R⟩—OCH₂CO—polyacrylamide

a

Fmoc(NHCHRⁱCO)ₙOCH₂—⟨benzene ring with R⟩—OCH₂CO—polyacrylamide

b | c

H(NHCHRⁱCO)ₙOCH₂—⟨benzene ring with R⟩—OCH₂CO—polyacrylamide

Reagents: a, (FmocNHCHR¹CO)₂O

b, Morpholine

c, Active ester of FmocNHCHRⁱCO₂H (i = 2 to n)

Scheme 7.14. Solid-phase synthesis on polyacrylamide resin.

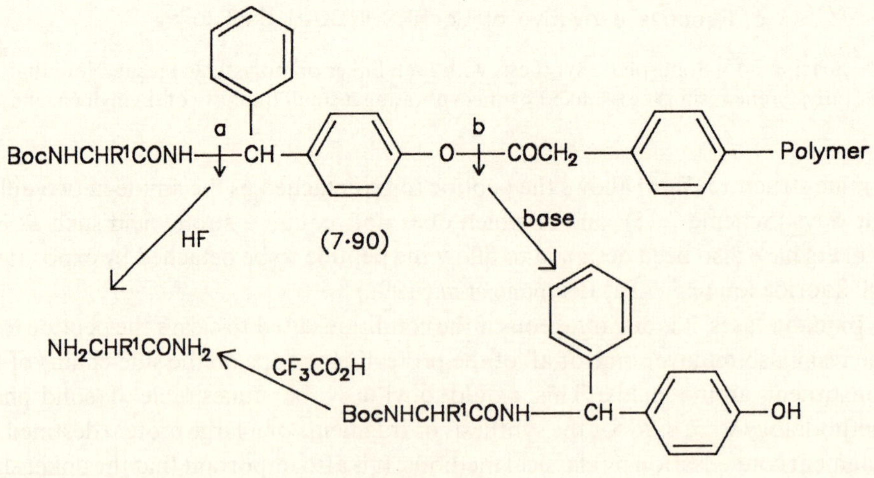

BocNHCHR¹CONH—CH—⟨phenyl⟩—O—COCH₂—⟨benzene ring⟩—Polymer

a   (7·90)   b

HF   base

NH₂CHR¹CONH₂ ← CF₃CO₂H ← BocNHCHR¹CONH—CH—⟨benzene ring⟩—OH

Scheme 7.15

160

FmocNHCHRCOO ... SiMe$_3$ ... CONHCH$_2$CONH —— Polymer

(7.91)

the growing peptide to fold back on the linker and/or resin or upon itself by non-covalent interactions. Spectrophotometric evidence has been obtained that indicates that β-structures can be formed in the assembly of certain peptides containing numerous bulky hydrophobic groups either as amino-acid side-chains or as protecting groups that are frequently difficult to obtain in good yield. A decapeptide sequence in acyl-carrier protein (residues 65–74) is notoriously difficult to obtain and its synthesis is used as a routine test of any new development in SPPS. The use of a polar solvent such as 1,1,1,3,3,3-hexafluoro-2-propanol can often accelerate difficult coupling steps. The best method, however, so far developed for preventing hydrogen-bonding participating in forming β-structures involves the temporary substitution of the nitrogen atom of selected peptide bonds. The 2-hydroxy-4-methoxybenzyl group (Hmb) has been developed to be used for this purpose in combination with protection of α-amino groups with the Fmoc group (Quibell *et al.*, 1994a, b; Johnson and Quibell, 1994). The 2-hydroxy group renders the group labile to acidolysis so that it can easily be removed when the peptide assembly has been completed. The Hmb group is thus orthogonal to the Fmoc group. In order to stabilise the Hmb group against premature removal during peptide assembly, the 2-hydroxy group is acetylated. The 2-hydroxy group is freed by 20% aqueous pyridine ready for acidolytic removal of Hmb groups. An additional advantage of the Hmb group is its tendency to increase the solubility of intermediates during peptide synthesis.

When the Fmoc group is used for routine protection of α-amino groups, the bond between the peptide and the linker can be designed to be quite sensitive to acidolysis. The Sheppard group has designed a series of suitable linkers by incorporating an electron-donating group into the benzyl ester moiety. A typical example is the group **7.92**. Detachment is also possible enzymatically using the linker **7.93**. Phosphodiesterase cleaves on either side of the phosphodiester group, but this is not a problem. If the *C*-terminal amino acid is required to have a free carboxy group, the substituted benzyl groups can be removed from the liberated peptide by hydrogenolysis or by strong acid or the peptide can be converted into a hydrazide ready for an azide coupling.

It is most important to avoid the formation of byproducts during the coupling reactions. Syntheses using unsymmetrical anhydrides that might cleave in the wrong

161

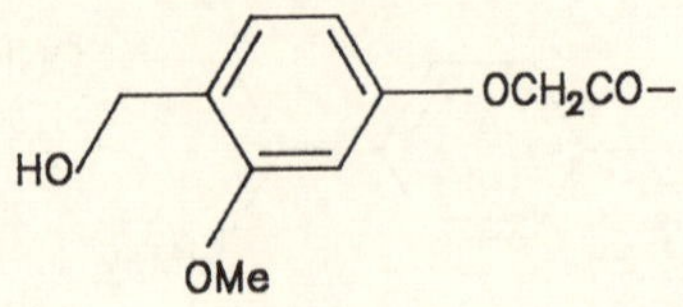

(7·92)

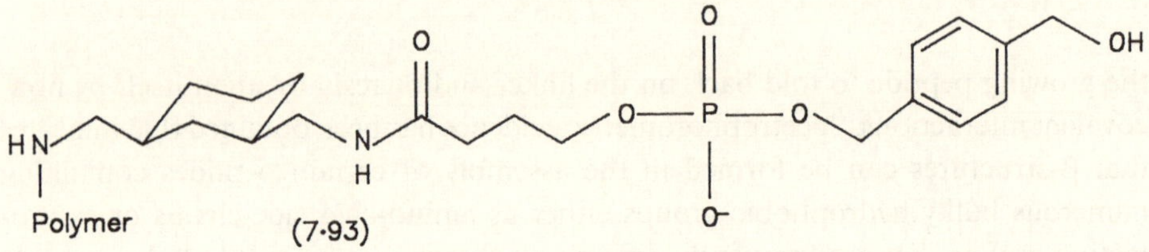

(7·93)

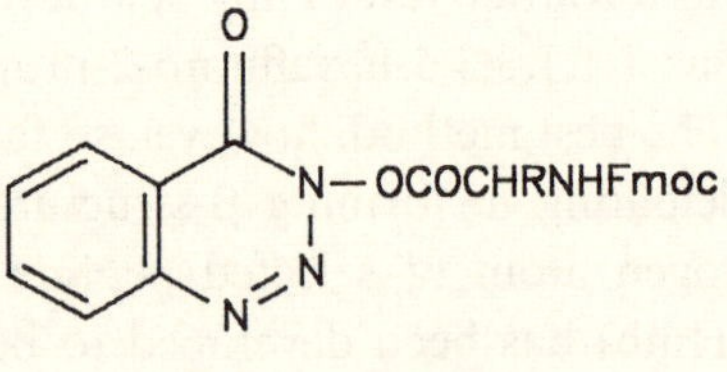

(7·94)

direction or carbodiimides that might yield *N*-acylureas have lost some of their popularity in SPPS. In spite of the intrinsic wastefulness associated with the use of symmetrical anhydrides, the absence of ambiguity in their reactions has increased their popularity even if sometimes their use is confined to the attachment of only the *C*-terminal residue to the linker. If reactive esters are used, they must afford rapid coupling in high yield with negligible enantiomerisation. Halogenated aryl (e.g. pentafluorophenyl) esters are commonly used for this purpose. Alternatively, esters of *N*-hydroxy-heterocycles are strongly favoured because of the low risk of enantiomerisation. The BOP reagent (**7.83**) has been mentioned above, but an alternative approach involves the use of esters of 3,4-dihydro-3-hydroxy-4-oxobenzotriazine (**7.94**) (Dhdt). The free hydroxy compound is a sensitive indicator for unreacted amine, since the anion absorbs strongly in the visible region (λ_{max}=440 nm). The resin is intensely yellow at the start of a coupling reaction and this fades as the reaction proceeds to completion. Photometric measurement of the colour of the resin thus provides a signal for feedback control of the computer controlling the hardware used for SPPS (Cameron *et al.*, 1987). An alternative and simpler method for following the acylation step in SPPS requires only the addition of a suitable acid–base indicator such as bromophenol blue (Krchňák *et al.*, 1988). This method has two advantages. The indicator is in solution so that, by using a continuous circulation mode, the progress of the reaction can be followed spectrophotometrically by recording the decrease in absorbance at 429 nm. In addition, any of the usual reactive esters can be used in the coupling step. The amount of indicator is small and the authors claim that no acylation of the phenolic hydroxy groups in the indicator by trans-esterification can be detected.

SPPS can be carried out manually with very simple laboratory glassware, but very esoteric hardware incorporating a computer to programme the addition of reagents, monitor the progress of reaction steps or simply to terminate steps after a predetermined interval is now available. An important technical advance in the design of mechanised SPPS provides for continuous flow or recirculation of reagents through the matrix on which the synthesis is taking place. This ensures that the utilisation of reagents and hence the yield are maximised. For details of this modification, see Atherton and Sheppard (1989). Various types of chemistry (Boc and Fmoc) are permissible, but one of the most important recent advances is the possibility of carrying out multiple syntheses simultaneously. The earliest method of effecting this involved the enclosure of batches of the resin on which syntheses were to be carried out in small bags. These were manually inserted into the reaction vessel if a particular amino acid was to be coupled to the peptides under assembly. Not surprisingly, this rapidly became known as the 'teabag' technique. Numerous ingenious methods for achieving the same end under computer control have since been devised. For example, microsyntheses of peptides can be carried out on plastic pins. This is a useful method for identifying the epitope regions in a protein. Every possible hexapeptide, heptapeptide and octapeptide in a polypeptide sequence can be synthesised, each on a separate pin, and tested for interaction with an antibody raised against the intact protein. Synthetic polymers are not mandatory as insoluble supports for SPPS. Microsyntheses can be carried out on pieces of filter paper. Despite the enormous technical progress that has been made, ingenuity shows no sign of drying up. Now that it is possible to monitor the progress of individual synthetic steps, it is clearly feasible to store these kinetic data and develop suitable software to interrogate this database to determine the optimum conditions for future syntheses. With the successful miniaturisation of SPPS, it is also possible to conceive of synthesising multiple peptides in safety in the kitchen or garage and posting the products to an appropriate laboratory for characterisation by mass spectrometry or biological testing. The development of combinatorial synthesis of peptides, especially in relation to pharmaceutical research, is discussed in Section 9.7.

7.10 Soluble-handle techniques

When SPPS was in its infancy, an alternative approach was examined. In this approach the same potential advantages of simple isolation and avoiding the need for full characterisation at each step were sought. The concept involved using a

$$MeO(CH_2CH_2O)_nH$$

(7.95)

'soluble handle' rather than an insoluble support for the growing peptide. For example, by using the 4-picolyl ester of the C-terminal amino acid, the growing peptide and handle are rendered soluble in acid and excess neutral reagents and byproducts can be extracted with organic solvents (Kisfaludy, 1979). Alternatively, the peptides can be asssembled on the free hydroxy group of the monomethyl ether of polyethylene glycol (7.95) (Mutter and Bayer, 1980). The C-terminal residue can be attached using the carbodiimide method of coupling in the presence of 1-N-hydroxybenzotriazole. Using polyethylene glycol of suitable average relative molecular mass, the polymer with its attached peptide is soluble in polar solvents for the coupling and deprotection stages, but can be precipitated by diethyl ether to provide a simple method of purification after the coupling of each amino acid. Despite the obvious attraction of such simple techniques, this methodology has not attracted the attention of manufacturers of scientific equipment.

7.11 Enzyme-catalysed peptide synthesis and partial synthesis

The biochemical role of proteolytic enzymes is to catalyse the hydrolysis of proteins and peptides either intra- or extra-cellularly and this is a process that is thermodynamically favoured:

$$R^1CONHR^2 + H_2O \rightleftharpoons R^1COOH + R^2NH_2.$$

One reason is that water is 55.5 M with respect to water, a concentration that cannot be physically matched for the other reactants. In addition, the carboxylic acid will tend to ionise and amine to acquire a proton in aqueous solution, which processes are incompatible with the reversal of the reaction. Nevertheless, before the mechanism of biosynthesis of proteins had been elucidated, it was generally thought that proteinases could effect protein synthesis by a simple reversal of the above reaction. A few model experiments gave credence to this belief. Thus, incubation of N-benzoylglycine and aniline in aqueous solution at about pH 5–6 with the enzyme papain, which is used as a meat tenderiser, gave quite good yields of N-benzoylglycine anilide as a crystalline precipitate. The synthetic success of this reaction depended on the insolubility of the product. Other attempted syntheses with different reactants and/or enzymes confirmed that the above reaction was a special case. When reaction was attempted in the presence of a high concentration of an organic solvent such as 1,4-butanediol, however, a moderate yield of synthetic peptide was formed. This was not so much due to the decrease in thermodynamic activity of water as to the increase in pK_a of the carboxylic acid. There was a small decrease in

pK_a of the amine in addition, but this was much less important. The combined effects of the changes in pK_a values increased the concentration of the neutral reactive species of acid and amine, an essential prerequisite for synthesis.

Despite the foregoing limitations, interest has returned to enzymic peptide synthesis under the biotechnological umbrella. One of the attractions, of course, is the high stereochemical specificity under many conditions and the absence of enantiomerisation in enzyme-catalysed syntheses. Synthesis can frequently be favoured by working in a two-phase system using a mixture of water and an immiscible organic solvent. Selective extraction of the product from the aqueous phase which contains the enzyme can afford very satisfactory yields. A variation on this theme uses a suitable detergent to form reverse micelles in a system containing mainly organic solvent with water limited to the interior of the micelles. Under some conditions, the stereospecificity of proteinases can be relaxed so that peptides containing D-amino acids can be synthesised. Consequently, if there is any risk that an intended reactant for peptide synthesis is not stereochemically pure, appropriate control experiments should be carried out.

An improved approach to enzyme-catalysed peptide synthesis stems from a thorough understanding of the kinetics and mechanism of action of proteinases. Many proteinases function by the Ping Pong Bi Bi mechanism (Roberts, 1977) and hydrolysis of an N-protected amino acid or peptide ester involves the acylation of a Ser or Cys side-chain by the ester with the liberation of the relevant alcohol or amino component and the formation of a covalent O- or S-acylated enzyme. The latter is hydrolysed in a second step:

$$R^1CONHCHR^2COR^3 + E\text{---}OH \text{ (or } E\text{---}SH) \rightarrow R^1CONHCHR^2CO\text{---}O(\text{or } S)\text{---}E$$
$$R^1CONHCHR^2CO\text{---}O(\text{or } S)\text{---}E + H_2O \rightarrow R^1CONHCHR^2COOH + E\text{---}OH \text{ (or }$$
$$E\text{---}SH).$$

When the enzyme is used to catalyse the synthesis of a peptide bond, the solvent is either non-aqueous or contains only a low concentration of water. In addition, of course, an amino component such as an amino acid or peptide ester replaces the water in the second step. Obviously, the amino component must be unprotonated for reaction to succeed. Synthesis is favoured over hydrolysis of the resultant peptide because an amide is kinetically a much worse substrate for a proteinase than is an ester. The rapid acylation of a proteinase by an N-protected amino acid or peptide aryl ester can be demonstrated experimentally using a stopped-flow apparatus with spectrophotometric facilities. A rapid burst of phenol is followed by steady-state release, showing that acylation of the enzyme is faster than hydrolysis of the acylated enzyme. No such burst is detectable if, for example, an N-acylated amino acid anilide is used as substrate. In fact, acylation is the rate-determining step with amide substrates.

It is customary to immobilise the enzyme on an insoluble support. This has two advantages. First, it conserves the enzyme and reaction can be carried out in a flow-through reactor. Secondly, an immobilised enzyme is much more resistant than is an enzyme in solution to denaturation by high concentrations of organic solvent and/or elevated temperatures used to accelerate a reaction. In complete contrast to the possibility of using elevated temperatures in order to accelerate enzyme-catalysed peptide synthesis, it has been discovered that yields of peptides and reaction rates can be increased by carrying out reactions in frozen aqueous solution (Hänsler and Jakubke, 1996). This observation could focus efforts in the future to find the best conditions for synthesis.

Before giving some examples of enzyme-catalysed synthesis of peptides, it is necessary to describe the specificity of a few typical proteinases. In many cases, most of the enzyme specificity is attributable to the side-chain of the amino acid in the substrate contributing the carbonyl group to the peptide bond that is cleaved. It is customary to label the amino-acid residues around the scissile bond as follows:

$$P_3 \text{—} P_2 \text{—} P_1 \text{—} P_1' \text{—} P_2' \text{—} P_3',$$

where the peptide bond between P_1 and P_1' is that for which the enzyme is predominantly specific. Thus, trypsin is specific for the cleavage or synthesis of peptides containing Arg or Lys as the P_1 residue. In contrast, chymotrypsin functions best if P_1 is an aromatic amino acid (e.g. Phe, Tyr or Trp) or an aliphatic hydrophobic amino acid (e.g. Leu or Met). Both enzymes fail to cleave a peptide bond in which Pro is at the P_1' position. Amino acids at nearby positions (sub-sites) may also play a minor role in quantitatively determining the kinetic sensitivity of a peptide bond to hydrolysis/synthesis by a proteinase. Some enzymes (exopeptidases) function either at the N-terminus (aminopeptidases) or at the C-terminus (carboxypeptidases), removing one residue at a time. Several enzymes of each type are known. One type of carboxypeptidase is specific for removing either aromatic amino acids or hydrophobic aliphatic amino acids. Another type of carboxypeptidase removes C-terminal residues of Lys and Arg. The biochemical complementarity of these enzymes and chymotrypsin and trypsin is easily appreciated. All these enzymes function in the small intestine and the carboxypeptidases effect further degradation of the fragments formed by the action of chymotrypsin and trypsin.

A few examples of enzyme-catalysed peptide synthesis will suffice to illustrate its power and flexibility. As indicated above, proteinases are usually stereospecific for amino-acid residues at positions P_1 and P_1'. Stereospecificity can sometimes be relaxed in solutions containing organic solvents. Thus papain affords quite high yields of protected dipeptides when it is incubated with Z—Gly—OEt and H—D—Ala—OEt, H—D—Leu—OMe, H—D—Phe—OMe or H—D—Val—OMe in aqueous methanol.

The next example has two interesting features. An intermediate for the synthesis of the sweetening agent 'Aspartame' has been synthesised from Z—Asp(OBzl)—OH

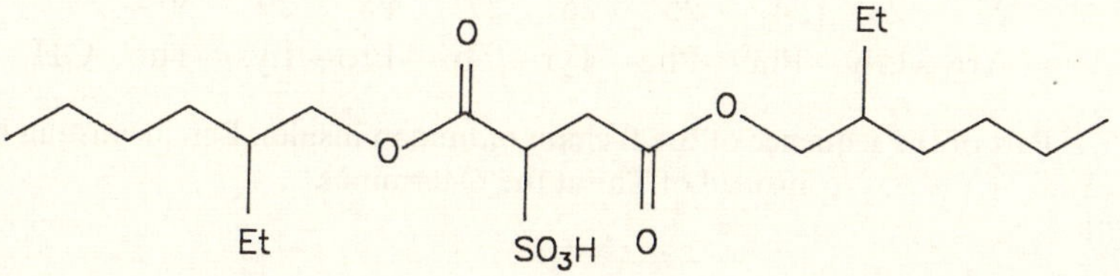

$$(7.96)$$

Ac—Ser(Bzl)—Tyr(Cl₂Bzl)—Ser(Bzl)—Met—Glu(OcHex)—His(Bom)—Phe—Arg(Tos)—OBzl

(assembled on oxime resin)

| HF

Ac—Ser—Tyr—Ser—Met—Glu—His—Phe—Arg—OH

| H—Trp(HCO)—Gly—Lys(ClZ)—Val—Pro—NH₂/trypsin
| CF₃CH(OH)CF₃/DMF (1:1) containing 4% H₂O

Ac—Ser—Tyr—Ser—Met—Glu—His—Phe—Arg—Trp(HCO)—Gly—Lys(ClZ)—Val—Pro—NH₂

| Deprotect

alpha—MSH

Scheme 7.16. Synthesis of α-melanocyte stimulating hormone.

and H—Phe—OMe in supercritical CO_2 using the proteinase thermolysin. Supercritical CO_2 is an excellent solvent for amino-acid and peptide derivatives and it does not denature the enzyme. Secondly, the synthesis of the peptide bond is accompanied by a decrease in volume so the reaction was carried out under a pressure of 300 atm, giving a 40% yield of Z—Asp(OBzl)—Phe—OMe.

As an example of the use of reverse micelles, chymotrypsin has been enclosed in reverse micelles formed with sodium bis-(2-ethylhexyl)sulphosuccinate (**7.96**) and used to catalyse the synthesis of Z—Ala—Phe—Leu—NH₂ from Z—Ala—Phe—OMe and H—Leu—NH₂ with isooctane as the bulk organic phase.

This section concludes with two examples of semi-synthesis catalysed by proteinases. The first case concerns the semi-synthesis of α-melanocyte stimulating hormone (α-MSH) from two fragments that had been made by chemical synthesis (Scheme 7.16). There are three interesting points to notice. One of the fragments was made by SPPS using a polystyrene resin containing a 4-nitrobenzophenone moiety,

22   23   24   25   26   27   28   29   30
—Arg—Gly—Phe—Phe—Tyr—Thr—Pro—Lys—Thr—OH

Figure 7.1. Part of the sequence of the B chain of human insulin. Porcine insulin has Ala instead of Thr at the C-terminus.

—$C_6H_4C(:NOH)C_6H_4NO_2$. The trypsin-catalysed coupling of the two fragments used an N-terminal peptide stripped of all side-chain protecting groups. Secondly, the C-terminal fragment contained a Lys residue, but this was unaffected by trypsin because its ε-amino group was protected. In the second example, human insulin, which differs from porcine insulin at only one locus (Figure 7.1), namely the C-terminal residue of the B chain, was semi-synthesised from the porcine protein by incubation with an excess of an ester of threonine in the presence of trypsin immobilised on spherical macroporous beaded cellulose.

Although the foregoing description of the use of proteinases in peptide synthesis reveals a lack of focus, there are signs that the situation is changing. The discovery that proteolytic enzymes not only survive dispersal in organic solvents with little or no water but also retain most of their catalytic activity, especially when they are immobilised on an inert support, accounts for there having been an increase in research along these lines. In addition, results from experiments with genetically engineered enzymes suggest that this may become a major field of research endeavour.

7.12 Cyclic peptides

7.12.1 Homodetic cyclic peptides

In a homodetic cyclic peptide, every pair of amino acids is joined by a conventional peptide bond. The simplest cyclodipeptides, 2,5-diketopiperazines, are derived from a dipeptide ester (Scheme 7.17). The ready formation of 2,5-diketopiperazines by a 6-*exo-trig* process results from the thermodynamic tendency to form stable six-membered rings even though the amide groups are in the less favoured *cis* rather than in the *trans* form. Cyclic peptides containing 3–5 amino-acid residues are formed much less readily on the whole and frequently the products of reaction contain linear oligopeptides and cyclic peptides containing twice the expected number of residues present in the parent linear peptide. The simplest method of making cyclic peptides involves preparing a reactive ester of an N-protected linear peptide, deprotecting the α-amino group, adding tertiary base and allowing cyclisation to occur in dilute solution to favour cyclisation over intermolecular condensation. The ease of cyclisation depends on the amino acids present and their configuration. The presence of amino acids such as Pro that favour β-turn formation facilitates cyclisation. The presence of one D-amino acid residue may achieve

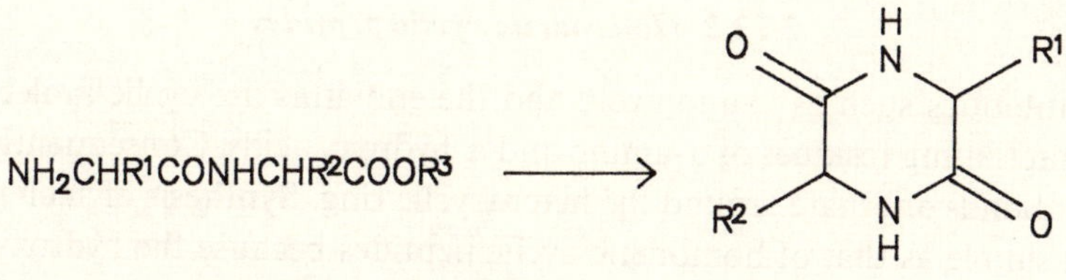

$$NH_2CHR^1CONHCHR^2COOR^3 \longrightarrow$$

Scheme 7.17.

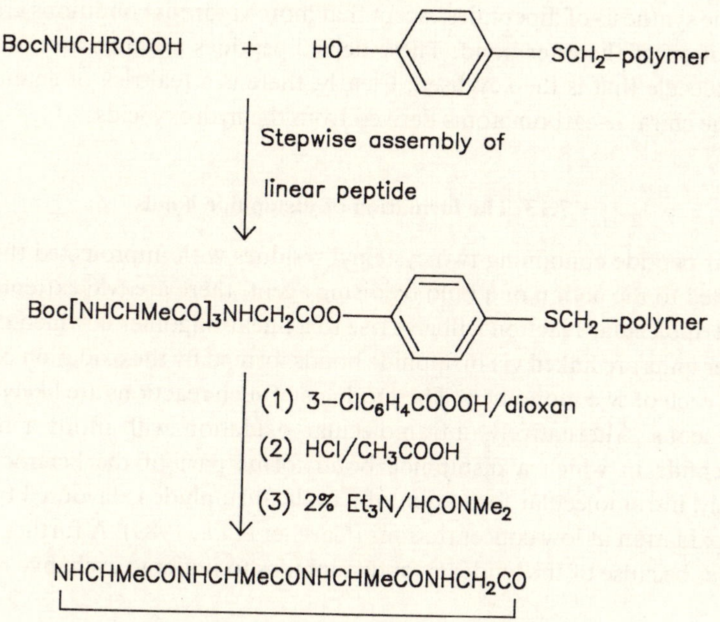

$$BocNHCHRCOOH \quad + \quad HO - \!\!\!\!\!\bigcirc\!\!\!\!\! - SCH_2\text{-}polymer$$

Stepwise assembly of

linear peptide

$$Boc[NHCHMeCO]_3NHCH_2COO - \!\!\!\!\!\bigcirc\!\!\!\!\! - SCH_2\text{-}polymer$$

(1) 3$-$ClC$_6$H$_4$COOOH/dioxan

(2) HCl/CH$_3$COOH

(3) 2% Et$_3$N/HCONMe$_2$

NHCHMeCONHCHMeCONHCHMeCONHCH$_2$CO

Scheme 7.18.

the same result. An alternative method of synthesising homodetic peptides involves SPPS on a resin that incorporates a safety catch in the linker moiety (Flanigan and Marshall, 1970) (Scheme 7.18). A linear peptide is assembled on a resin containing a 4-hydroxythiophenyl group. After deprotection of the N-terminal amino group, the thioether is oxidised to the sulphone which constitutes a reactive ester. Intramolecular cyclisation is favoured over the intermolecular formation of linear oligopeptides.

Cyclic peptides can be constructed by forming amide bonds using the side-chain carboxy groups of Asp and Glu and the ε-amino group of Lys. Such structures constrain the conformational freedom of the peptide main chain and protect the peptide from the action of proteinases such as trypsin and a proteolytic enzyme that cleaves glutamyl and aspartyl peptide bonds.

7.12.2 Heterodetic cyclic peptides

Some antibiotics such as valinomycin and the enniatins are cyclic molecules containing alternating residues of α-amino and α-hydroxy acids. Consequently, peptide and ester bonds alternate around the heterocyclic ring. Synthesis of such molecules is not as simple as that of homodetic cyclic peptides because the hydroxy group is considerably less nucleophilic than are amino groups. Commonly, ester building blocks such as $XNHCHR^1COOCHR^2COY$ are made by techniques similar to those used in the synthesis of dipeptides except that more vigorous conditions are required for formation of the ester bond. These depsidipeptides are then assembled into a linear molecule that is then cyclised. Clearly, there is a real risk of enantiomerisation of the chiral α-carbon atoms derived from the hydroxyacids.

7.13 The formation of disulphide bonds

If a linear peptide containing two cysteinyl residues with unprotected thiol groups is subjected to the action of a mild oxidising agent, there are two extreme possibilities. Intermolecular reaction will give rise to a linear oligomer in which the peptide monomer units are linked via disulphide bonds formed by the oxidation of one thiol group in each of two monomers. The products of such reactions are likely to be very heterogeneous. Alternatively, intramolecular oxidation will afford a heterodetic cyclic peptide in which a disulphide bond forms part of the heterocyclic ring. Obviously, intramolecular formation of a cyclic disulphide is favoured by carrying out the oxidation at low concentrations (Cavelier *et al.*, 1989). A further complication arises because of the tendency of disulphides to undergo exchange reactions:

$$R^1SH + R^2SSR^3 \rightleftharpoons R^1SSR^2 + R^3SH$$
$$R^1SH + R^2SSR^3 \rightleftharpoons R^1SSR^3 + R^2SH$$
$$R^1SH + R^1SSR^2 \rightleftharpoons R^1SSR^1 + R^2SH$$
$$R^2SH + R^1SSR^2 \rightleftharpoons R^2SSR^2 + R^1SH.$$

Only a catalytic amount of thiol is required to initiate this type of reaction sequence. Considering the first of these reactions, the position of equilibrium will be towards the right-hand side if the pK_a of R^3SH is less than the pK_a of R^1SH. The techniques used to synthesise linear and cyclic disulphides, especially if there are more than two cysteine residues, make use of orthogonal thiol protecting groups (Section 7.5) and employ methods for their regioselective introduction and removal, methods of oxidation of pairs of thiol groups and avoidance or even harnessing of the kind of exchange reactions described above. If a pair of thiol groups is protected by Acm groups, treatment with mild oxidising agents such as I_2, $Tl(CF_3COO)_3$ and CNI removes the blocking groups and simultaneously forms disulphide bonds (Section 7.5). If S-trityl groups are used, they are similarly removed with the formation of a

peptide$_1$—SH + Y—S—S—X—polymer ⟶ peptide$_1$—S—S—X—polymer +YSH

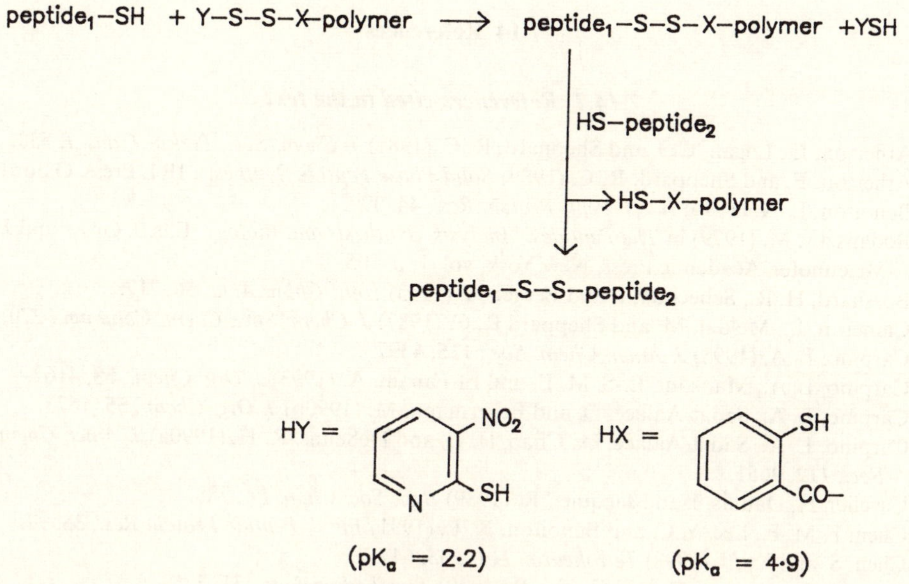

HS—peptide$_2$

⟶HS—X—polymer

peptide$_1$—S—S—peptide$_2$

HY =   (NO$_2$, SH on pyridine ring, N)   HX =   (SH, CO— on benzene ring)

(pK$_a$ = 2·2)   (pK$_a$ = 4·9)

Scheme 7.19.

disulphide bond. The relative rates of reaction, however, are dependent on the solvent. Thus a mixture of peptide$_1$—Cys(Trt)—peptide$_2$ and peptide$_3$—Cys(Acm)—peptide$_4$ treated with I$_2$ gave nearly pure

peptide$_1$—Cys—peptide$_2$
peptide$_1$—Cys—peptide$_2$

in CHCl$_3$ and CH$_2$Cl$_2$, mainly the unsymmetrical disulphide in MeOH and the alternative symmetrical disulphide in CHONMe$_2$. Several S-protecting groups can easily be removed with MeSiCl$_3$ or SiCl$_4$ in CF$_3$COOH and oxidation of the liberated thiol groups can be achieved with MeSOMe. The use of MeSOMe as an oxidising agent is preferable to oxidation in air because a higher concentration of oxidant can be used, thereby diminishing the possibility of disulphide-exchange reactions.

Although a considerable amount of work remains to be done, it is reasonable to expect to be able to form regiospecific disulphide bonds when more than two appropriately S-protected cysteine residues are available, either intramolecularly or intermolecularly, using suitable methods of deprotection and oxidation. Thus, if a peptide containing two S-protected Cys residues is assembled by SPPS, blocking groups may be removed and intramolecular oxidative formation of disulphide bonds carried out before detachment of the peptide from the resin. Another potentially regiospecific method uses solid-phase methodology and the disulphide-exchange reactions described above (Scheme 7.19). The desired reactions are favoured by the relative acidities of the thiols involved. The pK$_a$ of the thiol group in peptides containing cysteine is >10.

7.14 References

7.14.1 References cited in the text

Atherton, E., Logan, C. J. and Sheppard, R. C. (1981) *J. Chem. Soc., Perkin Trans. I*, 538.

Atherton, E. and Sheppard, R. C. (1989) *Solid Phase Peptide Synthesis*, IRL Press, Oxford.

Benoiton, L. N. (1994) *Int. J. Pept. Protein Res.*, **44**, 399.

Bodanszky, M. (1979) in *The Peptides: Analysis, Synthesis and Biology*, Eds. J. Gross and J. Meienhofer, Academic Press, New York, vol. 1, p. 105.

Bosshard, H. R., Schechter, I. and Berger, A. (1973) *Helv. Chim. Acta*, **56**, 717.

Cameron, L., Meldal, M. and Sheppard R. C. (1987) *J. Chem. Soc., Chem. Commun.*, 270.

Carpino, L. A. (1993) *J. Amer. Chem. Soc.*, **115**, 4397.

Carpino, L. A., Mansour, E.-S. M. E. and El-Faham, A. (1993) *J. Org. Chem.*, **58**, 4162.

Carpino, L. A., Sadat-Aalaee, D. and Beyermann, M. (1990b) *J. Org. Chem.*, **55**, 1673.

Carpino, L. A., Sadat-Aalaee, D., Chao, H. G. and DeSelms, R. H. (1990a) *J. Amer. Chem. Soc.*, **112**, 9651.

Cavelier, F., Daunis, J. and Jacquier, R. (1989) *Bull. Soc. Chim. Fr.*, 788.

Chen, F. M. F., Lee, Y. C. and Benoiton, N. L. (1991) *Int. J. Peptide Protein Res.*, **38**, 97.

Chen, S. and Xu, J. (1992) *Tetrahedron Lett.*, **33**, 647.

Coste, J., Le-Nguyen, D. and Castro, B. (1990) *Tetrahedron Lett.*, **31**, 205.

Flanigan, E. and Marshall, G. R. (1970) *Tetrahedron Lett.*, 2403.

Hänsler, M. and Jakubke, H. D. (1996) *Amino Acids*, **11**, 379.

Hiskey, R. G. (1981) in *The Peptides: Analysis, Synthesis and Biology*, Eds. J. Gross and J. Meienhofer, Academic Press, New York, vol. 3, p. 137.

Izumiya, N. and Muraoka, M. (1969) *J. Amer. Chem. Soc.*, **91**, 2391.

Johnson, T. and Quibell, M. (1994) *Tetrahedron Lett.*, **35**, 463.

Jones, J. H. (1979) in *The Peptides: Analysis, Synthesis and Biology*, Eds. J. Gross and J. Meienhofer, Academic Press, New York, vol. 1, p. 65.

Jones, J. (1994) *The Chemical Synthesis of Peptides*, Clarendon Press, Oxford.

Kemp, D. S. (1979) in *The Peptides: Analysis, Synthesis and Biology*, Eds. J. Gross and J. Meienhofer, Academic Press, New York, vol. 2, p. 417.

Kisfaludy, L. (1979) in *The Peptides: Analysis, Synthesis and Biology*, Eds. J. Gross and J. Meienhofer, Academic Press, New York, vol. 2, p. 417.

Knorr, R., Trzeciak, A., Bannwarth, W. and Gillessen, D. (1989) *Tetrahedron Lett.*, **30**, 1927.

Krchňák, V., Vágner, J., Šafář, P. and Lebl, M. (1988) *Coll. Czech. Chem. Commun.*, **53**, 2542.

Meienhofer, J. (1979a) in *The Peptides: Analysis, Synthesis and Biology*, Eds. J. Gross and J. Meienhofer, Academic Press, New York, vol. 1, p. 197.

Meienhofer, J. (1979b) in *The Peptides: Analysis, Synthesis and Biology*, Eds. J. Gross and J. Meienhofer, Academic Press, New York, vol. 1, p. 263.

Merrifield, B. (1986) *Science*, **232**, 341.

Milton, R. C. deL., Milton, S. C. F. and Kent, S. B. H. (1992) *Science*, **256**, 1445.

Mutter, M. and Bayer, E. (1980) in *The Peptides: Analysis, Synthesis and Biology*, Eds. J. Gross and J. Meienhofer, Academic Press, New York, vol. 2, p. 285.

Quibell, M., Turnell, W. G. and Johnson, T. (1994a) *Tetrahedron Lett.*, **35**, 2237.

Quibell, M., Turnell, W. G. and Johnson, T. (1994b) *J. Org. Chem.*, **59**, 1745.

Ramage, R., Barron, C. A., Bielecki, S., Holden, R. and Thomas, D. W. (1992) *Tetrahedron*, **48**, 499.

Ramage, R., Hopton, D., Parrott, M. J., Richardson, R. S., Kenner, G. W. and Moore, G. A. (1985) *J. Chem. Soc., Perkin Trans. I*, 461.

Rich, D. H. and Singh, J. (1979) in *The Peptides: Analysis, Synthesis and Biology*, Eds. J. Gross and J. Meienhofer, Academic Press, New York, vol. 1, p. 241.

Roberts, D. V. (1977) in *Enzyme Kinetics*, Cambridge University Press, Cambridge.

Savrda, J. and Wakselman, M. (1992) *J. Chem. Soc., Chem. Commun.*, 812.

Stewart, J. M. (1981) in *The Peptides: Analysis, Synthesis and Biology*, Eds. J. Gross and J. Meienhofer, Academic Press, New York, vol. 3, p. 169.

Stewart, J. M. and Young, J. D. (1984) in *Solid Phase Peptide Synthesis*, 2nd ed., Pierce Chemical Co.

Tam, J. P., DiMarchi, R. D. and Merrifield, R. B. (1981) *Tetrahedron Lett.*, **22**, 2851.

Veber, D. F., Milkowski, J. D., Varga, S. L., Denkewalter, R. G. and Hirschmann, R. (1972) *J. Amer. Chem. Soc.*, **94**, 5456.

7.14.2 References for background reading

Bailey, P. D. (1990) *An Introduction to Peptide Chemistry*, John Wiley & Sons, New York.

Bodanszky, M. (1993) *Principles of Peptide Synthesis*, 2nd ed., Springer-Verlag, Berlin.

Grant, G. A. (Ed.) (1992) *Synthetic Peptides: A User's Guide*, W. H. Freeman & Co., New York.

Kocieński, P. J. (1994) *Protecting Groups*, Georg Thieme Verlag, Stuttgart.

Merrifield, B. (1993) *Life During a Golden Age of Peptide Synthesis. The Concept and Development of Solid Phase Peptide Synthesis*, American Chemical Society, Washington.

8

Biological roles of amino acids and peptides

8.1 Introduction

Amino acids fulfil three broad classes of function in biology. They serve as building blocks in prokaryotes and plant and animal eukaryotes for the synthesis of peptides and proteins. Most peptides derive from the processing of proteins, but some such as glutathione, folate and peptide antibiotics are biosynthesised by specific non-ribosomal routes (see Chapter 9). In contrast, particular amino acids, especially glycine, are required in the synthesis of a wide variety of small molecules, including alkaloids, purine and pyrimidine nucleotides, porphyrins, creatine and phospho-creatine. The second role of amino acids is to act as intermediates in incorporating or disposing of small molecules. For example, arginine is involved in various reaction sequences in the disposal of unwanted nitrogen as urea and the production of perhaps the most unexpected biomolecule, nitric oxide. Again, methionine makes its *S*-methyl group available for methylation reactions via the intermediate *S*-adenosylmethionine. Finally, some important biomolecules are derived by the metabolism of amino acids. Enzymic decarboxylation of some of the coded amino acids or of a hydroxylated derivative gives rise to important cellular messengers and hormones. Alternatively, an amino acid and an α-keto acid can undergo a trans-amination reaction and, since several α-keto acids are important metabolic inter-mediates, this reaction offers a simple route to some of the inessential amino acids. The amino group can also be removed oxidatively from an amino acid, giving rise to an α-keto acid. Some amino acids such as histidine and tryptophan undergo unique ring-opening reactions that lead, through rather complex pathways, to glu-tamic acid and alanine, respectively.

Let us consider for a moment two points from the above paragraph. The process of ribosomal protein synthesis, which involves deoxyribonucleic acids (DNAs) that make up the genes and ribonucleic acids that (i) convey the genetic information from the nucleus to the ribosomes and (ii) transfer amino acids to the polypeptide under assembly, and the regulation of protein synthesis are subjects big enough for a large textbook. On the other hand, the metabolism of the amino acids that result from the breakdown of proteins involves small molecules but is complicated by the diversity of metabolic paths. This latter panoply of reactions can be complicated by a variety of inborn errors of metabolism. Clearly, such a catholic range of metabolic sequences and cross-connections with the metabolism of other foodstuffs such as carbohydrates and fats rules out any detailed treatment of the biological roles of amino acids. All that can be done in a text of this size is to mention briefly some points that may already have been encountered by students of biochemistry and to provide some signposts to textbooks for chemists who wish or need to cross the divide that frequently and regrettably separates them from a study of the chemistry that produces and maintains life.

8.2 The role of amino acids in protein biosynthesis

The blueprint for synthesising a protein is stored within the coding DNA (cDNA) of the genes in the chromosomes. In order to encode the information to incorporate one of the twenty amino acids likely to be present in a protein, one or two purine or pyrimidine bases are not sufficient since there would be too few unique combinations of bases. It has been proved conclusively that each amino-acid residue in a protein is encoded by a triad or codon of purine/pyrimidine bases in the gene. The sequences of bases in the cDNA and messenger RNA (mRNA) are complementary, with the important exception that some sections of cDNA are not transcribed into RNA. Thus, the bases A, C, G and T in cDNA become U, G, C and A in mRNA. The sections of DNA that are transcribed are known as *exons*, whereas untranscribed sections are known as *introns*.

Because there are sixty-four possible base triads or codons for only twenty amino acids and a stop codon to indicate when transcription should stop, there is considerable redundancy in the genetic code (Table 8.1). The twenty amino acids do not occur with equal frequency and it is notable that the commonest amino acids are encoded by several base triads. Moreover, variation of base sequence for a particular amino acid is most permissible in the third position of a codon. Thus GGA, GGC, GGG and GGU all code for glycine. Obviously, there is a good chance that a mutation in the gene will not lead to a change in the amino-acid sequence of the resultant protein. Conversely, a rarer amino acid has fewer possible codons. Thus, Met and Trp have only one codon each. There is no strict correlation, however, between the number of codons for an amino acid and the frequency of its occurrence in proteins. For example, Arg occurs less commonly than does the other strongly basic amino acid, Lys, yet there are six codons for Arg and only two for Lys. Apart from the common redundancy in the third base of a codon, it is notable that the second base can often specify homology of amino acids encoded. Thus, all the triads with U as the middle base code for hydrophobic amino acids. Again,

Table 8.1. *The genetic code*

First base	Second base			
	U	C	A	G
U	UUU Phe	UCU Ser	UAU Tyr	UGU Cys
	UUC Phe	UCC Ser	UAC Tyr	UGC Cys
	UUA Leu	UCA Ser	UAA Stop	UGA Stop
	UUG Leu	UCG Ser	UAG Stop	UGG Trp
C	CUU Leu	CCU Pro	CAU His	CGU Arg
	CUC Leu	CCC Pro	CAC His	CGC Arg
	CUA Leu	CCA Pro	CAA Gln	CGA Arg
	CUG Leu	CCG Pro	CAG Gln	CGG Arg
A	AUU Ile	ACU Thr	AAU Asn	AGU Ser
	AUC Ile	ACC Thr	AAC Asn	AGC Ser
	AUA Ile	ACA Thr	AAA Lys	AGA Arg
	AUG Met[a]	ACG Thr	AAG Lys	AGG Arg
G	GUU Val	GCU Ala	GAU Asp	GGU Gly
	GUC Val	GCC Ala	GAC Asp	GGC Gly
	GUA Val	GCA Ala	GAA Glu	GGA Gly
	GUG Val	GCG Ala	GAG Glu	GGG Gly

Note:
[a] AUG is part of the initiation signal for translation but also codes for internal Met residues.

codons that start with UC and AC specify hydroxyamino acids and codons that start with GA encode dibasic amino acids. To summarise, the genetic code is highly degenerate and non-random.

For some considerable time, it was thought that the genetic code was universal, especially because genetic-engineering experiments showed repeatedly that eukaryotic genes could be expressed in bacteria such as *E. coli*. More recently, however, it has been found that mitochondria have their own genetic code and protein-synthetic machinery. This has led to discussions about the evolutionary origin of mitochondria, a topic that cannot be pursued here.

The synthesis of a protein requires the mRNA as a template containing the full sequence of codons, including the codon to terminate synthesis. The ribosomes, which orchestrate protein synthesis, read the mRNA in the $5' \rightarrow 3'$ direction. (The $5'$ end has a phosphate group on the $5'$-carbon atom of a ribose moiety whereas the $3'$ end has a phospate group on the $3'$-carbon atom of ribose). Protein biosynthesis requires a transfer ribonucleic acid (tRNA) to convey an amino acid to the growing peptide chain. tRNAs are specific for each codon and contain 60–95 nucleotides, a few of which have unusual structures. The $3'$ end of the tRNA has the sequence

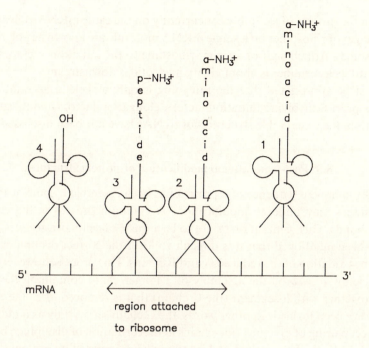

1 Loaded tRNA ready to attach to mRNA

2 Loaded tRNA attached to mRNA codon

3 tRNA bearing peptide fragment to be

   transferred to tRNA2

4 Unloaded tRNA leaving mRNA

Scheme 8.1.

—CCA and the amino acid to be introduced into the protein esterifies the 3'-hydroxy group. At a point about halfway along the sequence of the tRNA there is a triad of purine/pyrimidine bases (an anticodon) that is complementary to the codon for the amino acid to be introduced. This ensures that the tRNA binds non-covalently to the mRNA at the correct place. The process of polypeptide elongation is depicted in Scheme 8.1. The amino group of an aminoacyl-tRNA is believed to attack the ester carbonyl group of the adjacent peptidyl-tRNA and the whole of the peptide chain that has been assembled up to that point is transferred to the aminoacyl-tRNA. The empty tRNA that has given up its peptide chain dissociates and the mRNA and ribosome engages the next aminoacyl-tRNA. Note that peptide assembly proceeds from the *N*-terminus to the *C*-terminus, the opposite of the recommended laboratory practice (Chapter 7). This demonstrates the tight stereochemical control exerted in enzyme-catalysed reactions. Finally, several polypeptide chains

177

can be in the process of assembly concurrently on the same mRNA. These multiple assemblages of ribosomes on a single mRNA molecule are known as polyribosomes or polysomes. Attachment of another ribosome to the initiation site does not take place until its forerunner is about eighty nucleotides downstream.

It must be appreciated that the foregoing is only a skeletal account of a very complex process involving initiation factors, elongation factors and release factors. In addition, the remarkable structures of tRNAs have not been discussed here.

8.3 Post-translational modification of protein structures

When the complete sequence of a protein has been assembled, some of the amino-acid residues may undergo modification. A common process is the cleavage of peptide bonds. This seems at first sight to be a waste of cellular resources, but it has already been mentioned that it is difficult to form the correct disulphide bonds in insulin by oxidation of the reduced A and B chains. This process, however, is carried out efficiently *in vivo* by the assembly of a longer chain containing the A and B chains together with a segment (the C chain) that is removed *after* the disulphide bonds have been formed. In other words, the C chain serves only as a molecular jig for correct pairing of cysteinyl side-chains in the formation of disulphide bonds. The C chain is subsequently excised and its amino acids recycled by proteolysis.

Another example of a peptide sequence in a protein forcing it into a biologically useful conformation is found with collagen. This consists of a triple helix with chains of more than 1000 amino-acid residues, many of which are post-translationally modified. The latter steps, consisting *inter alia* of hydroxylation of Pro and Lys residues and 5-hydroxylysine residues, occur before the triple helix is formed, because the enzymes involved do not act on the helical structure. When the individual peptide chains of collagen are synthesised, there are *N*- and *C*-terminal sequences each containing about 100 amino-acid residues. These sequences favour the formation of a triple helix. When this has been achieved, the terminal sequences are removed. The sequences of these temporary terminal sequences are quite different from the main body of the collagen monomers, which consists of triads of the type Gly—X—Y, where X and Y are often proline or 3- or 4-hydroxyproline.

Post-translational modifications often involve changes that confer important biological properties. The phosphorylation of side-chain hydroxy groups is a good example. The modification of a single Ser residue regulates the activity of some enzymes. For example, glycogen phosphorylase mobilises glucose-1-phosphate for energy-producing metabolism of glycogen stored in the liver and muscle. A molecule of glucose-1-phosphate is produced when one glucose residue is detached by phosphorolysis of an $\alpha(1 \rightarrow 4)$ glycosidic linkage. Glycogen phosphorylase is a dimeric protein consisting of two identical chains (97 kD). It can undergo reversible phosphorylation (Scheme 8.2) and exhibits allosteric behaviour (a change of conformation) in the presence of molecules required for or produced by glycolysis.

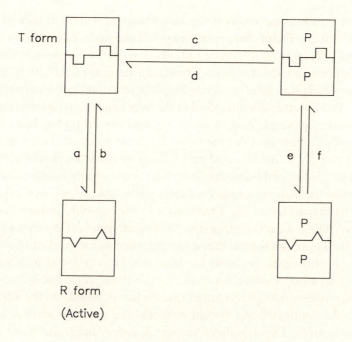

T form

P
P

a | b

e | f

R form

(Active)

P

P

Reagents: a, AMP: b, ATP and/or G_6P

c, ATP/phosphorylase kinase

d, Phosphoprotein phosphatase

e, spontaneous

f, Glucose

Scheme 8.2.

The catalytically active R form (phosphorylase b) is not phosphorylated. Binding of ATP or of glucose-6-phosphate produces an allosteric conversion into the inactive, unphosphorylated T form, a process that is reversed by competitive binding of AMP. The latter is a product of the breakdown of ATP and signals that more glycogen should be broken down to fuel glycolysis. The production of ATP and glucose-6-phosphate by glycolysis regulates this process by switching the R form to the T form. The active site of the T form is sterically inaccessible to the substrate. ATP also phosphorylates a single residue in each chain of phosphorylase b, giving phosphorylase a, a process catalysed by phosphorylase kinase. This process can be reversed by a separate enzyme, phosphoprotein phosphatase. The phosphorylase is also regulated by reversible phosphorylation.

Another example of post-translational modification that is important in confer-

179

ring activity on an enzyme involves the conversion of some Glu side-chains in certain blood-clotting factors into γ-carboxyglutamic acid. This process requires vitamin K and CO_2 under oxidising conditions. Several pro-enzymes in the blood-clotting mechanism undergo this form of post-translational modification. Since the blood-clotting mechanism involves a complex set of reactions, only one example will be given here. Prothrombin is synthesised in the liver and it contains ten residues of a modified form of glutamic acid in which a γ-carboxy group has been inserted. These ten residues occur in a domain near the N-terminus of the molecule. The modified Glu resembles malonic acid and undergoes decarboxylation very easily, especially under acidic conditions. Prothrombin is enzymically inactive and is converted into active thrombin in a rather complex proteolytic hydrolysis mediated, for example, by the enzyme factor Xa. Thrombin is a rather specific proteolytic enzyme and cleaves two different peptides from the N-terminal region of the six-unit protein fibrinogen. Fibrinogen is a soluble blood protein whereas the product of the action of thrombin on fibrinogen, fibrin, is the insoluble protein found in blood clots. Because injury to a major blood vessel can be life-threatening, the blood-clotting mechanism has to respond rapidly to minimise the loss of blood. On the other hand, any defect in the control of this system enhances the risk of a serious coronary thrombosis or a stroke. Deprivation of vitamin K supply limits the extent of post-translational modification of the Glu residues in prothrombin and this causes the activation in response to thrombin to be markedly inhibited. Much of our supply of vitamin K is produced by the synthetic efforts of intestinal bacteria so that the required dietary intake of the vitamin, while important, is lower than it otherwise would be. Neonatal infants require some time to establish an adequate population of intestinal flora, so they are at risk in any situation requiring mobilisation of the blood-clotting mechanism. In contrast, patients who have suffered and survived a coronary thrombosis are routinely administered warfarin (**8.1**), which is a competitive inhibitor of vitamin K (**8.2**) and therefore limits the extent of post-translational modification of Glu residues in prothrombin. Periodic determination of the clotting time of the blood of such patients allows rather fine control of the blood-clotting mechanism.

Proteins may be acetylated, usually on the α-amino group's N-terminal residue but occcasionally on the ε-amino group of a lysine residue. Acetylation is effected with acetylcoenzyme A (**8.3**), although other biochemical acetylating agents are well known. α-Melanotropin or α-melanocyte-stimulating hormone (α-MSH) provides an example of post-translational acetylation:

$$\text{Ac—Ser—Tyr—Ser—Met—Glu—His—Phe—Arg—Trp—Gly—Lys—Pro—}$$
$$\text{Val—NH}_2.$$

A more complicated example is afforded by histone H4 from calf thymus. The N-terminal serine residue is N-acetylated and may be O-phosphorylated. ε-N-

180

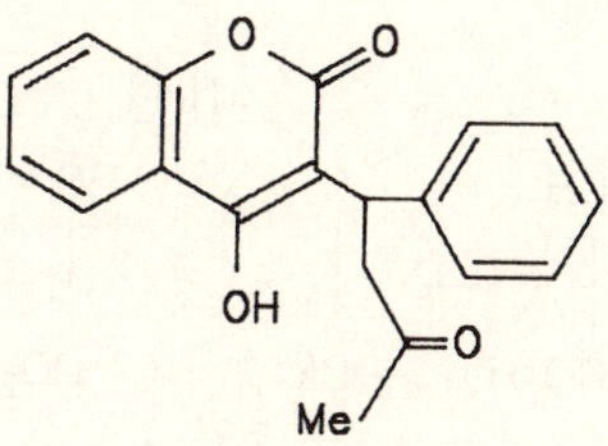

(8·1)

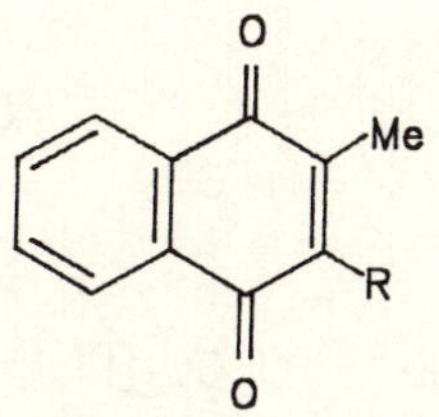

(Vitamin K_1, R = $C_{20}H_{38}$

Vitamin K_2, R = $C_{40}H_{73}$

Vitamin K_3, R = H)

(8·2)

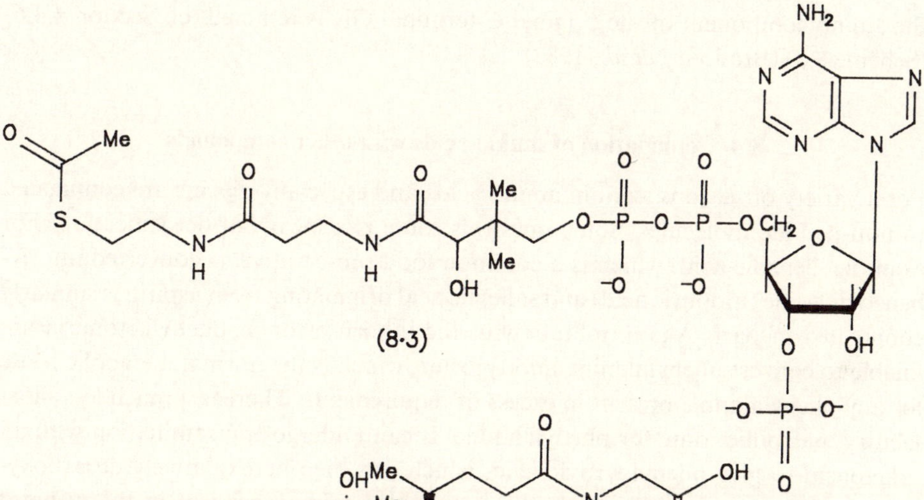

(8·3)

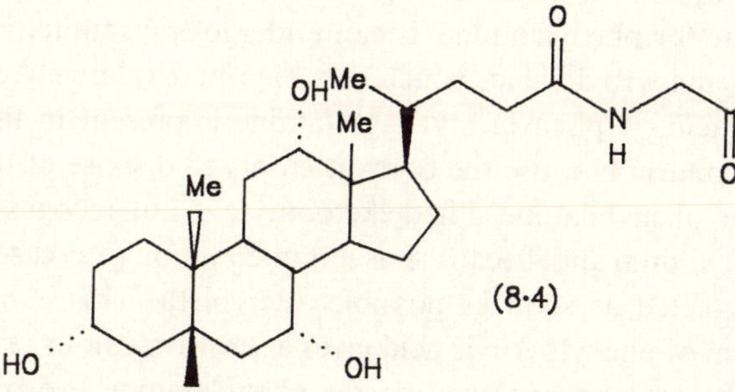

(8·4)

Acetylation can also affect Lys^5, Lys^8 and Lys^{12}. The ε-amino group of Lys^{20} is also modified, but by mono- or di-methylation. The biochemical methylating agent is S-adenosylmethionine (Section 8.5). The biosyntheses of coenzyme A and S-adenosylmethionine provide additional examples of the biochemical utilisation of cysteine and methionine, respectively.

The final example of post-translational modification to be considered deals with the conversion of the C-terminal amino acid into a primary amide. The obvious

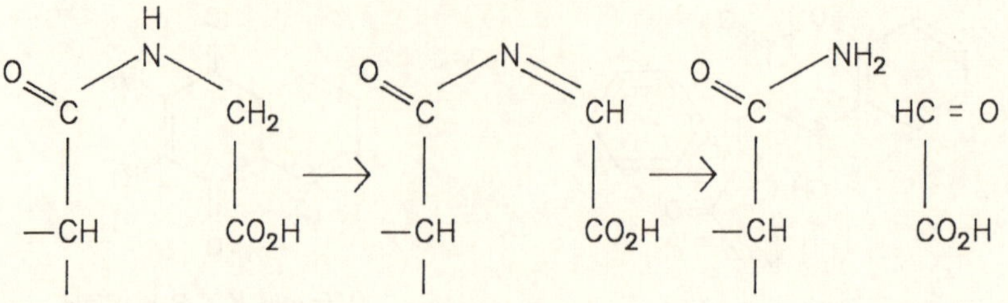

Scheme 8.3.

route involving coupling ammonia to a peptide is not employed; neither is there a codon for NH_3. Instead a peptide, usually with a *C*-terminal glycine residue, is enzymically oxidised to an enamine derivative that is then hydrolysed so that only the amino component of the original *C*-terminal Gly is retained (cf. Section 4.4.6) (Scheme 8.3) (Bradbury *et al.*, 1982).

8.4 Conjugation of amino acids with other compounds

For a variety of reasons, certain amino acids and especially glycine are conjugated to non-peptide molecules. Sometimes this route is a means of detoxification. For example, benzoic acid, which is a common food preservative, is converted into *N*-benzoylglycine (hippuric acid) and salicylic acid originating from aspirin is similarly conjugated in part. Again, infants with the inborn error of phenylketonuria are unable to convert phenylalanine into tyrosine, which is the normal metabolic route for any phenylalanine present in excess of requirements. There is a much less satisfactory metabolic route for phenylalanine. It can undergo transamination with an α-ketoacid to give phenylpyruvic acid, which can then be oxidatively decarboxylated to phenylacetic acid. *N*-phenylacetyl-L-glutamine is present in the urine of people with phenylketonuria because the body attempts to dispose of the unusual metabolic products of phenylalanine. Phenylketonuria, if not recognised, causes mental retardation of children and, because it is not uncommon (one case per 10 000 births), it must be detected as soon as possible after birth. This is not difficult because the enolic form of phenylpyruvic acid gives a sensitive colour reaction with $FeCl_3$. Insertion of the missing gene coding for phenylalanine hydroxylase into patients is a prime target for gene therapy. It should be noted that, since tyrosine cannot be formed by sufferers from phenylketonuria, tyrosine is a dietary essential amino acid for these people. Conversely, the content of phenylalanine in the diet must be very carefully controlled, especially during pregnancy if the mother has the genetic error. There must be enough for the synthesis of essential proteins, but not enough to compromise normal intellectual development. A shortage of tyrosine can limit melanin synthesis and so people with phenylketonuria tend to have a fair complexion and blond hair.

Conjugation involving amino acids is a normal metabolic route in some cases. For example, a bile acid such as cholic acid is partly converted into glycocholic acid (**8.4**) in man, possibly to increase the critical micellar concentration in water and thereby facilitate the transport of lipids around the body.

8.5 Other examples of synthetic uses of amino acids

Creatine (**8.5**) is present in striated muscle and is synthesised from glycine in two stages. First, the guanidino group of arginine is transferred to glycine to give guanidinoacetic acid (glycocyamine). Then the nitrogen atom nearest to the carboxy group is methylated by *S*-adenosylmethionine. The guanidino group is phosphorylated catalytically by phosphocreatine kinase and this molecule is available for supporting muscular activity over a limited period. Phosphocreatine slowly cyclises with loss of inorganic phosphate to give creatinine (**8.6**). Creatinine is excreted normally in urine so creatine must be synthesised continuously. Creatine is excreted in significant amounts in normal health only by menstruating women and this results from the breakdown of smooth muscle cells of the endometrium. It is also excreted by people suffering from pathological conditions involving muscle wasting, such as muscular dystrophy and thyrotoxicosis.

Glycine contributes C^4, C^5 and N^7 in the biosynthesis of purine ribonucleotides whereas the side-chain amide group of glutamine contributes N^3 and N^9. The initial purine ribonucleotide synthesised is inosine-5′-phosphate (**8.7**). Production of adenosine-5′-phosphate (**8.8**) uses aspartic acid to convert the 6-carbonyl group into the 6-amino group. In the synthesis of guanosine-5′-phosphate (**8.9**), inosine-5′-phosphate is first oxidised to xanthosine-5′-phosphate (**8.10**) and then glutamine contributes its amide nitrogen to furnish the 6-amino group. The biosynthesis of pyrimidine ribonucleotides differs rather remarkably from the assembly of the purine analogues. In purine nucleotide biosynthesis, the heterocyclic ring system is assembled in a stepwise fashion on to C^1 of 5-phosphoribose-1-pyrophosphate. In pyrimidine ribonucleotide biosynthesis, however, the ribose-5′-phosphate moiety is not mobilised until the pyrimidine ring has been assembled. Pyrimidine ribonucleotide synthesis proceeds through orotic acid (**8.11**), which is assembled through (i) carbamoyl phosphate whose NH_2 group is derived from glutamine, (ii) *N*-carbamoyl aspartate produced by the interaction of carbamoyl phosphate and aspartic acid, (iii) ring-closure of *N*-carbamoyl aspartate to dihydroorotate and (iv) oxidation of the latter to orotic acid. In summary, N^1 of the pyrimidine ring stems from glutamine and N^3, C^4, C^5 and C^6 are contributed by aspartic acid. After reaction with 5-phosphoribose-1-pyrophosphate the orotidine-5′-phosphate is decarboxylated to give uridine-5′-phosphate. Unlike the cases of adenosine-5′- and guanosine-5′-phosphate, for which the amino groups are directly obtainable from the inosine and xanthosine derivatives, uridine-5′-phosphate is not directly converted into cytidine-5′-phosphate. Instead, uridine-5′-phosphate is first converted

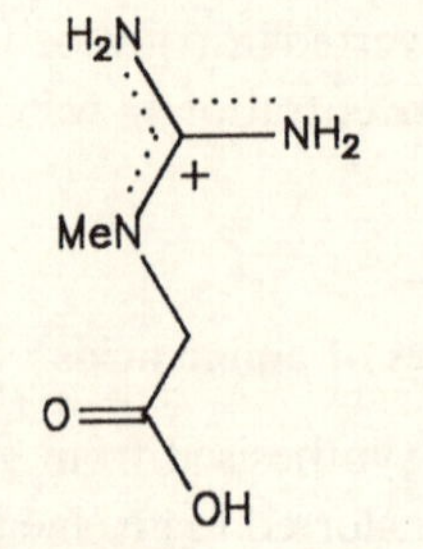

(8·5)

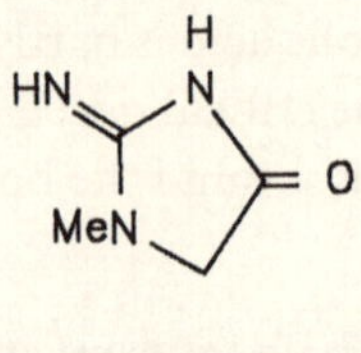

(8·6)

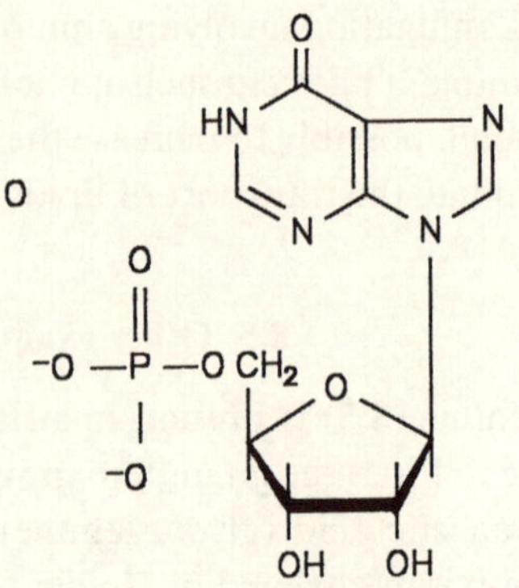

(8·7)

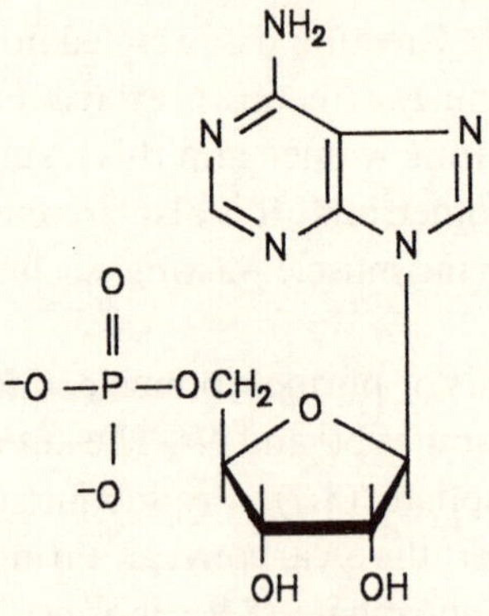

(8·8)

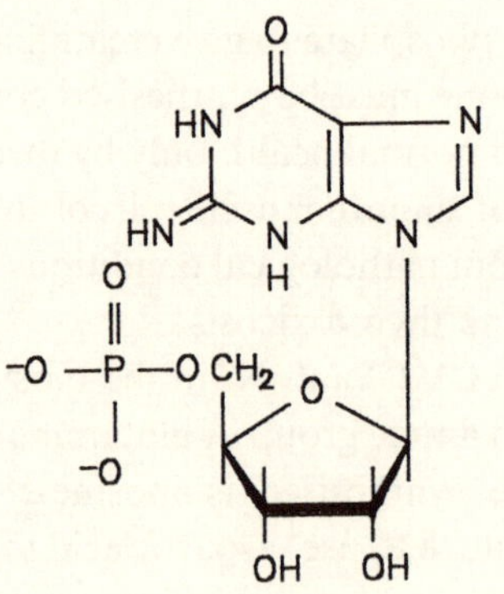

(8·9)

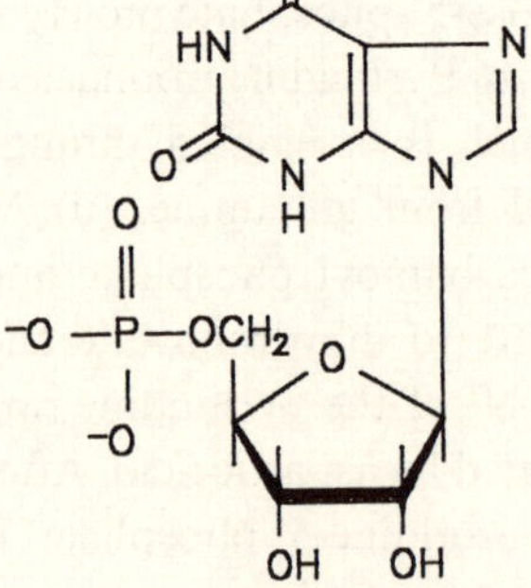

(8·10)

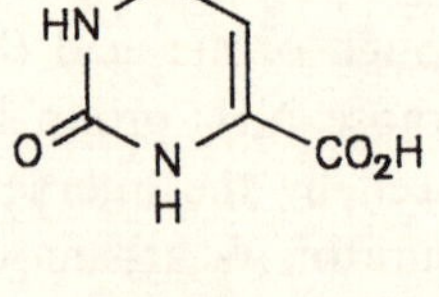

(8·11)

into the 5′-diphosphate and then the 5′-triphosphate. The latter reacts with ATP and NH_3 to give cytidine-5′-triphosphate and this is converted into the monophosphate.

Porphyrins and the haem pigment of haemoglobin are assembled using glycine and succinoyl-CoA to produce δ-aminolaevulinic acid as an intermediate. (Note that the latter is commonly abbreviated to ALA which must not be confused with Ala, the abbreviation for alanine). All four nitrogen atoms in the porphyrin ring are derived from glycine. Several genetic defects in the biosynthesis of porphyrins are known collectively as porphyrias. The commonest of these conditions, acute intermittent porphyria, has probably had an important influence on world history. It is an autosomal dominant disease caused by a deficiency of an enzyme system, uroporphyrinogen I synthase and cosynthase, and affects Laplanders in particular. The British King George III (who reigned from 1760 until 1820 when his son became Regent) suffered from this all his adult life and he had repeated episodes of serious psychological derangement. The intransigent attitude of George III and his government towards the British colonies in North America led inexorably to the American Revolution and the independence of the colonies which later became the United States of America. How different world history might have been but for a genetic defect in porphyrin biosynthesis in one man!

Except in the case of malnutrition of various types, the intake of nitrogen exceeds that required for the biosynthesis of proteins and nucleic acids *inter alia* and the body must dispose of the excess. Although some ammonia is produced and some purines are excreted as uric acid, the basic properties of the former and the insolubility of the latter indicates the need for some other route. Urea is an ideal waste product because it is neutral and very soluble. The synthesis of urea from ammonia and bicarbonate involves the amino acids arginine, ornithine, citrulline and aspartic acid and consists of a cycle of reactions discovered by Krebs and Henseleit before the former emigrated from Nazi Germany. The cycle is shown in Scheme 8.4. Carbamoyl phosphate (**8.12**), produced by the interaction of ATP, bicarbonate and ammonia, carbamoylates ornithine (**8.13**) to form citrulline (**8.14**). This reaction is essentially irreversible and constitutes the rate-determining step for the cycle. The carbamoyl group is converted into a substituted guanidino group by reaction with the α-amino group of aspartic acid, thereby producing argininosuccinate (**8.15**). This is next broken down to arginine and fumarate. The arginine is then hydrolysed by arginase to urea, which is excreted via the kidneys, and ornithine, which is ready for another turn of the cycle. The fumarate can be reconverted into aspartic acid for further production of urea via a separate cycle involving the formation of malate, then oxaloacetate and finally aspartic acid. The urea cycle occurs in the liver, the major detoxification organ of the body. The synthesis of carbamoyl phosphate and its reaction with ornithine occurs in mitochondria whereas the remainder of the cycle occurs in the cytosol of cells. Consequently, ornithine requires a specific transport system to enter the mitochondria and citrulline must be exported therefrom.

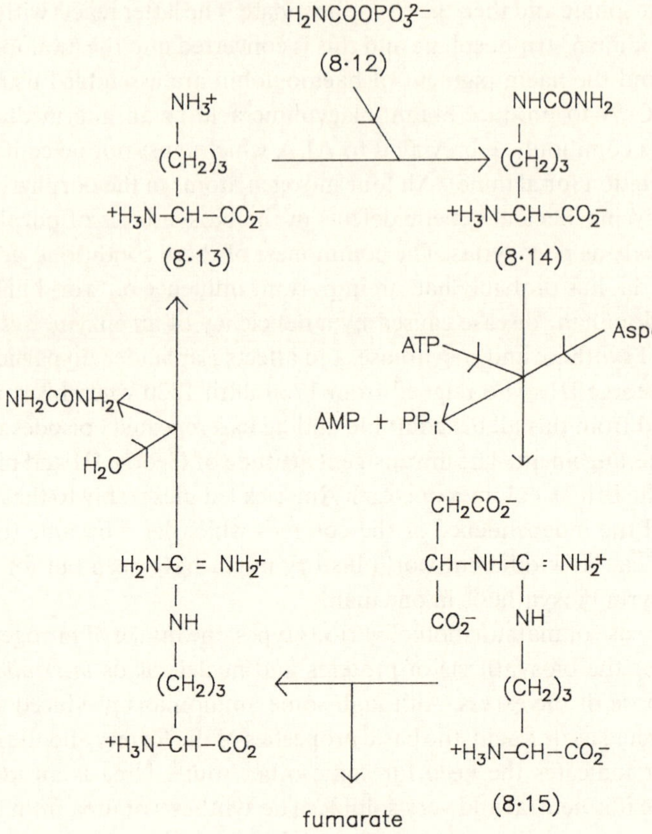

$$H_2NCOOPO_3^{2-}$$

(8·12)

$$\begin{array}{ccc} NH_3^+ & & NHCONH_2 \\ | & & | \\ (CH_2)_3 & \longrightarrow & (CH_2)_3 \\ | & & | \\ {}^+H_3N-CH-CO_2^- & & {}^+H_3N-CH-CO_2^- \end{array}$$

(8·13)   (8·14)

ATP ⤬ Aspartate

AMP + PP$_i$

$$NH_2CONH_2$$
$$H_2O$$

$$\begin{array}{cc} H_2N-\overset{}{C} = NH_2^+ & CH_2CO_2^- \\ | & | \\ NH & CH-NH-C \vdots NH_2^+ \\ | & | \quad | \\ (CH_2)_3 & CO_2^- \quad NH \\ | & | \\ {}^+H_3N-CH-CO_2 & (CH_2)_3 \\ & | \\ & {}^+H_3N-CH-CO_2^- \end{array}$$

fumarate   (8·15)

Scheme 8.4.

Before 1987, nitric oxide was regarded solely as an atmospheric pollutant that formed nitrous and nitric acids in the presence of moist air. It was known that nitric oxide, like carbon monoxide, bound to haemoglobin with a considerably higher affinity than did oxygen. These data were consistent with the view that nitric oxide was poisonous. It was with considerable surprise that it was found not only that nitric oxide was present in a wide range of organisms but also that it was actually synthesised not only in the cardiovascular system, for example, but also in the brain. It is a powerful vasodilator and the efficacy of glyceryl trinitrate, which has been used for over a century for treating angina pectoris, is due to the formation of nitric oxide as a metabolite. Nitric oxide is a messenger molecule. In the central nervous system, it influences the release of glutamate, which is a neurotransmitter; whereas in peripheral tissues it behaves as a non-adrenergic, non-cholinergic nerve transmitter. As indicated above, it helps to control vascular relaxation and hence blood flow and other smooth muscles associated with the gastrointestinal tract; operation of the bladder sphincter and erection of the penis occur under its influence. It modulates platelet aggregation, limiting the risk of a thrombus forming at the site of a vascular injury. It also has an antimicrobial effect on some extremely pathogenic micro-organisms. Nitric oxide also kills tumour cells. Nevertheless, it is not just a straightforward molecular Santa Claus. Overproduction of nitric oxide may well be involved in several pathological states, such as septicaemia, rheumatoid arthritis, osteoarthritis and graft-versus-host disease. Clearly, in good health there must be homeostatic mechanisms to control the production of nitric oxide. In pathological situations in which nitric oxide appears to be overproduced, synthetic inhibitors of NO synthases are required as drugs and this is an active area of research.

The formation of nitric oxide, like that of urea, involves arginine and its conversion into citrulline, but the resemblance is more apparent than real. Nitric oxide synthases require O_2, reduced nicotinamide adenine dinucleotide phosphate (NADPH), flavin adenine dinucleotide (FAD), flavin mononucleotide (FMN), haem, tetrahydrobiopterin and Ca^{2+} ions as well as arginine, although NO synthase from macrophages and liver contains tightly bound calmodulin, which holds Ca^{2+} ions. The overall method of production of NO is clearly a complex process; it involves a five-electron oxidation of one $-NH_2$ group present in the guanidino group of arginine and proceeds through N-hydroxyarginine (Scheme 8.5). For a detailed review of the biochemistry of nitric oxide, see Kerwin *et al.* (1995).

8.6 Important products of amino-acid metabolism

Enzymic decarboxylation of amino acids produces an array of biochemically active amines. Glutamate decarboxylase produces γ-aminobutyric acid (GABA), a neurotransmitter with inhibitory rather than excitatory properties. Histidine decarboxylase produces histamine, which has several functions. It activates a $(H^+–K^+)$-ATPase in the gastric mucosa that effects secretion of 0.15 M HCl for digestion of protein by pepsin and also to prevent the growth of a prolific array of bacteria in the stomach. Secondly, it is released from special storage cells (mast cells) in an inflammatory response. At first sight, it might be judged an evolutionary misfortune to cause this painful condition, but the possibility of thrombosis or gangrene would be far worse. The inflammatory response ensures that the injured part is irrigated by fluid, permitting the entry of leukocytes to deal with invading micro-organisms. Finally, histamine is released in anaphylaxis, namely the consequence of re-exposure to a substance to which the body has become allergic. The consequent drop in blood pressure caused by histamine can induce anaphylactic shock, which can be rapidly fatal.

Tyrosine is hydroxylated by tyrosine hydroxylase before decarboxylation, giving 3,4-dihydroxyphenylalanine (DOPA) (Scheme 8.6). The decarboxylation then gives dopamine, which functions as an inhibitory neurotransmitter in particular parts of the brain such as the substantia nigra. Degeneration of this region occurs in Parkinson's disease. Administration of dopamine is ineffective because it cannot

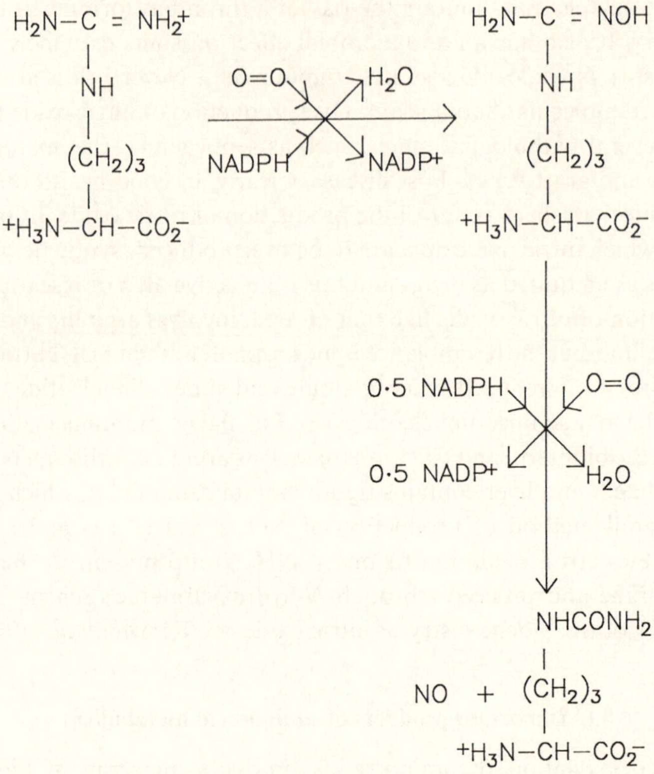

$$H_2N-C = NH_2^+$$
$$|$$
$$NH$$
$$|$$
$$(CH_2)_3$$
$$|$$
$$^+H_3N-CH-CO_2^-$$

$O=O \quad H_2O$

$NADPH \quad NADP^+$

$$H_2N-C = NOH$$
$$|$$
$$NH$$
$$|$$
$$(CH_2)_3$$
$$|$$
$$^+H_3N-CH-CO_2^-$$

$0.5\ NADPH \qquad O=O$

$0.5\ NADP^+ \qquad H_2O$

$$NHCONH_2$$
$$|$$
$$NO\ +\ (CH_2)_3$$
$$|$$
$$^+H_3N-CH-CO_2^-$$

Scheme 8.5.

cross the blood–brain barrier. DOPA, however, can reach this target and presumably then undergoes decarboxylation. This treatment can alleviate the condition. In other tissues, metabolism of tyrosine does not stop at dopamine. The latter is oxidised to noradrenaline (**8.16**) and then methylated to adrenaline (**8.17**) and the latter is the major product. Both noradrenaline and adrenaline are sympathetic neurotransmitters and are produced by the adrenals. The effects produced by these compounds are quite dramatic and adrenaline has been called the 'fight or flight hormone'. The physiological effects depend on the receptor at which the hormone arrives, but some of the more notable effects include acceleration of the heart rate (tachycardia) and an increase in breakdown of glycogen, effects that would assist both in fighting and in fleeing.

Like tyrosine, tryptophan is hydroxylated before decarboxylation (Scheme 8.7) to give serotonin (**8.18**). Serotonin is released from platelets during aggregation following vascular injury. The serotonin causes vasoconstriction, decreasing the rate of blood loss. Serotonin is also present in the pineal gland, which is located between the two hemispheres of the brain. In the pineal gland, it is enzymically *N*-acetylated and then *O*-methylated to give melatonin (**8.19**). The rate of synthesis of melatonin

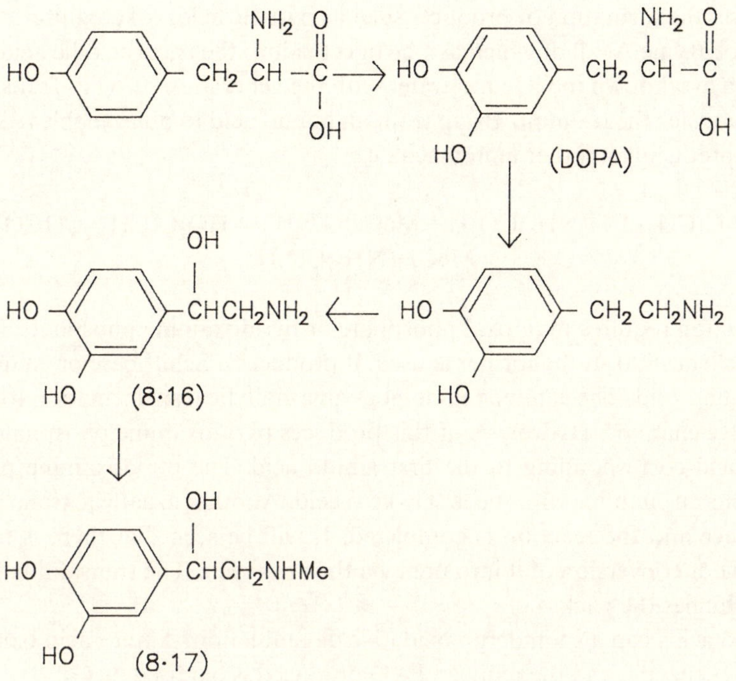

Scheme 8.6.

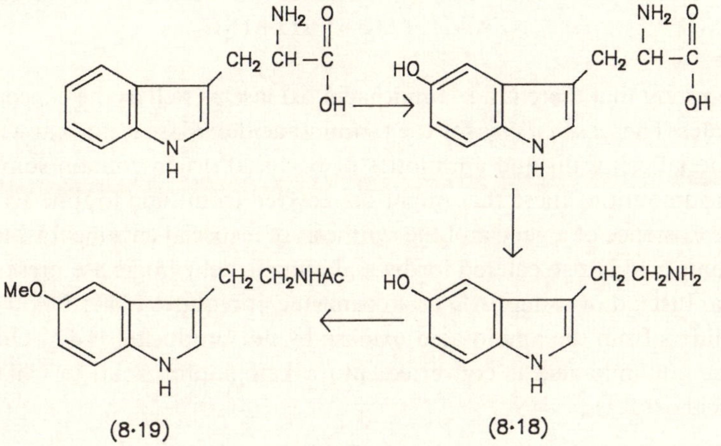

Scheme 8.7.

from serotonin is regulated by the intensity of light to which the animal is exposed. The *O*-methylation step decelerates in bright light. Melatonin causes the ovaries of female rats to be smaller than normal and also reduces the output of luteinising hormone from the pituitary.

An alternative metabolic route for amino acids involves removal of the α-amino group and the formation of products such as pyruvic acid, α-ketoglutaric acid and acetyl-coenzyme A, all of which can be injected into the tricarboxylic acid cycle for complete breakdown to CO_2 and water with the generation of ATP. Transaminases simply transfer the α-amino group from an amino acid to an available α-keto acid, thereby producing another amino acid, e.g.

$$HO_2C(CH_2)_2CH(NH_2)CO_2H + MeCOCO_2H \rightleftharpoons HO_2C(CH_2)_2COCO_2H + MeCH(NH_2)CO_2H.$$

This reaction requires pyridoxal phosphate or pyridoxamine phosphate as a coenzyme (Scheme 8.8). If the former is used, it produces a Schiff base or enamine with the α-amino acid. The enamine undergoes an azaallylic transformation to form the alternative enamine. Hydrolysis of this produces pyridoxamine phosphate and the α-keto acid corresponding to the first amino acid. The pyridoxamine phosphate now forms an enamine with the first α-keto acid. Another azaallylic transformation takes place and the reaction is completed. It will be seen that there is no loss of ammonia or conversion of it into urea via the urea cycle. The transamination route simply shuffles the pack.

Amino acids can also undergo oxidative de-amination. Amino-acid oxidases use FAD as coenzyme and the reduced FAD produced is oxidised by O_2:

$$RCH(NH_2)CO_2H + FAD + H_2O \rightarrow RCOCO_2H + NH_3 + FADH_2$$
$$FADH_2 + O_2 \rightarrow FAD + H_2O_2.$$

It is interesting that there is a D-amino-acid oxidase as well as the expected one for L substrates. The *raison d'être* for the D-amino-acid oxidase is not known. Granted that bacterial cell walls and antibiotics of bacterial origin contain some D-amino acids, the amount of these that might be recycled from time to time scarcely warrants the existence of a gene and the synthesis of a special enzyme for the purpose.

Glutamic acid is also catered for by a glutamate dehydrogenase present in mitochondria. Instead of using FAD as a coenzyme, it requires either NAD or NADP. It also differs from the amino-acid oxidase by not producing H_2O_2. On the other hand, the glutamic acid is converted into α-ketoglutaric acid, just like with the amino acid oxidases.

8.7 Glutathione

Glutathione (**8.20**) γ-glutamylcysteinylglycine, is biosynthesised in a stepwise manner from the *N*-terminus but not by the ribosomal route. Instead, each peptide bond is formed under the control of a specific enzyme, glutamyl-cysteine synthetase and glutathione synthetase, respectively, with ATP as a substrate to form an acyl phosphate as an unsymmetrical anhydride. Reduced glutathione (GSH) with a thiol

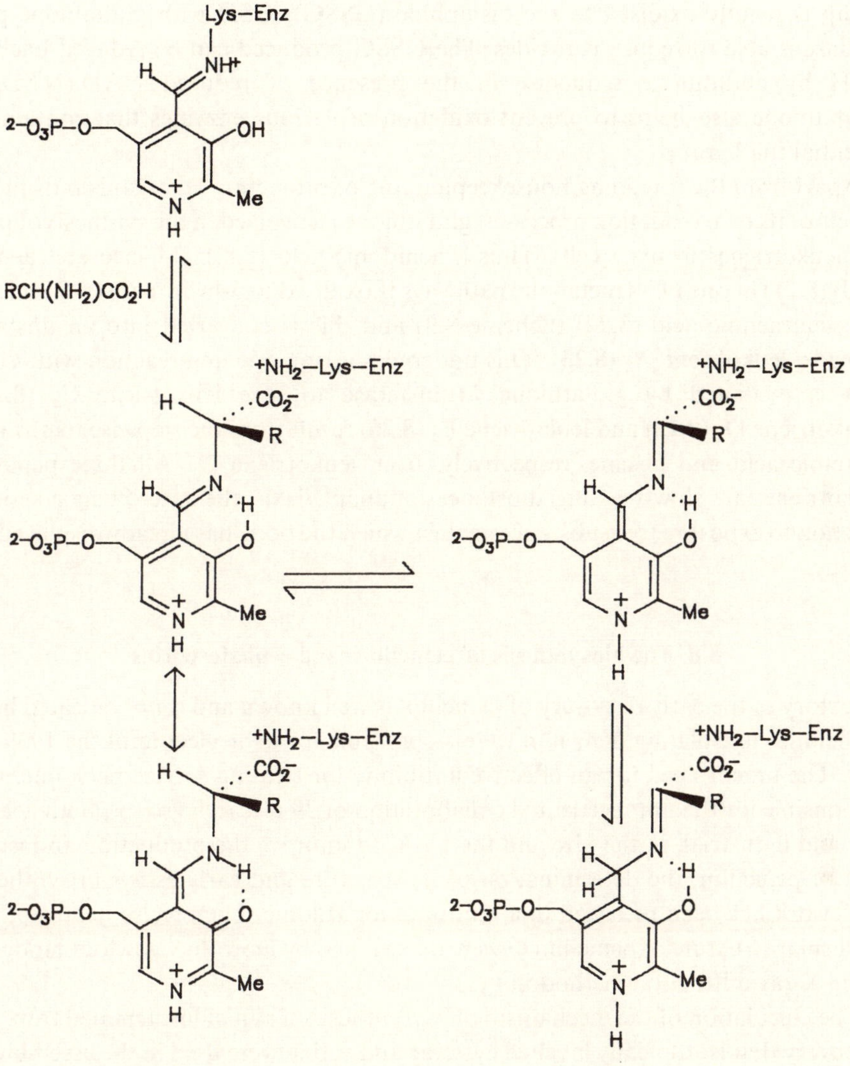

(1) Hydrolysis of the ketimine structure, K, gives $RCOCO_2H$
    and pyridoxamine phosphate

(2) Exchange reaction of K with $R'COCO_2H$ and reversal of the
    above reactions gives $R'CH(NH_2)CO_2H$ and pyridoxal phosphate

Scheme 8.8.

191

group is readily oxidised to the disulphide (GSSG). GSH with glutathione per-oxidase is able to reduce peroxides. The GSSG produced can be reduced back to GSH by glutathione reductase in the presence of reduced NAD (NADH). Glutathione also helps to prevent oxidation of various enzymes that contain an essential thiol group.

Apart from the foregoing housekeeping role of protecting an organism from the effects of its own oxidation processes, glutathione is involved in the synthesis of pep-tidoleukotrienes in mast cells. Thus arachidonic acid (5,8,11,14–eicosatetraenoic acid) **(8.21)** in one of its metabolic pathways is oxidised to 5-hydroperoxy-6,8,11,14-eicosatetraenoic acid **(8.22)** (Scheme 8.9) and this is converted into an unstable epoxide, leukotriene A_4 **(8.23)**. This undergoes a ring-opening reaction with GSH that is mediated by glutathione-S-transferase to give leukotriene C_4 **(8.24)**. Leukotriene D_4 **(8.25)** and leukotriene E_4 **(8.26)** result from the stepwise removal of glutamic acid and glycine, respectively, from leukotriene C_4. All three peptido-leukotrienes are slow-reacting substances of anaphylaxis, the latter being a serious reaction to exposure to a substance against which the body has already displayed an allergy.

8.8 The biosynthesis of penicillins and cephalosporins

The story of the early discovery of penicillin is well known and is not repeated here. Difficulties in isolating penicillin were responsible for its neglect until the 1939–45 war. The urgent need for an effective antibiotic for treating service personnel was responsible for the unprecedented collaboration of 39 research groups, both acade-mic and industrial, in the UK and the USA, to improve the production and isola-tion of penicillin, the determination of its structure and early essays in synthesis. This work has been recorded in a multi-author volume. The precise details of the molecular structure of penicillin G were determined by Dorothy Crowfoot Hodgkin using X-ray diffraction methodology.

The elucidation of the mechanism of biosynthesis of penicillin stemmed from the discovery that isotopically labelled cysteine and valine were used in the assembly of penicillin by *Penicillium chrysogenum* (Arnstein and Grant, 1954; Arnstein and Clubb, 1957). Cysteine and valine together with α-aminoadipic acid are used by *Cephalosporium acremonium* to synthesise penicillin N **(8.27)** and cephalosporin C **(8.28)**. Evidence was accumulated that a tripeptide, δ-(L-α-aminoadipoyl)-L-cysteinyl-D-valine (ACV) was formed as an intermediate. Since this tripeptide is not transported into mycelial cells, it must be synthesised intracellularly and synthesis of penicillin from the isotopically labelled tripeptide was demonstrated using a cell-free system. Clearly, ACV is not produced by a ribosomal synthesis of a protein fol-lowed by proteolytic processing. The enzyme involved, ACV synthetase, not only forms the two peptide bonds but also epimerises the valine residue. Thus, incuba-tion of [2-^{2}H]-valine with purified ACV synthetase completely removed deuterium

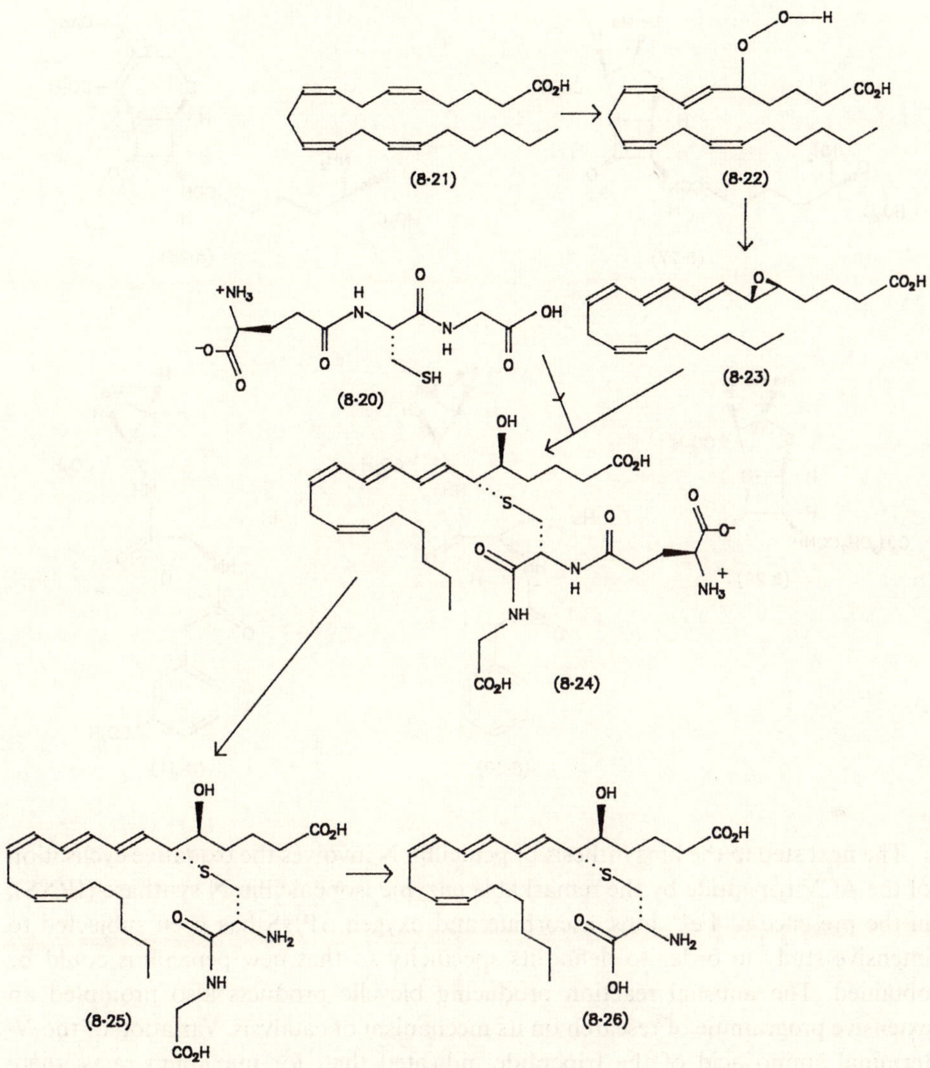

Scheme 8.9.

from C$_2$. *In vitro* synthesis using ^{18}O-valine resulted in the exchange of one oxygen atom, indicating that a valinyl enzyme is formed and that this is not fully reversible. This is consistent with the requirement for Mg^{2+} ions and 3 moles of ATP for the biosynthesis of penicillin N from the appropriate amino acids. The complexity of the synthesis of the tripeptide intermediate is underlined by the requirement for phosphopantothenic acid. Less surprising is the suggestion that the action of the enzyme ACV synthetase is the rate-limiting step in the biosynthesis of penicillin and cephalosporin.

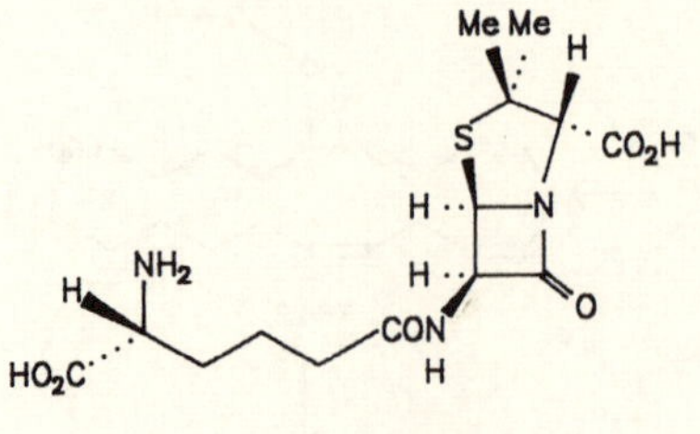

(8·27)

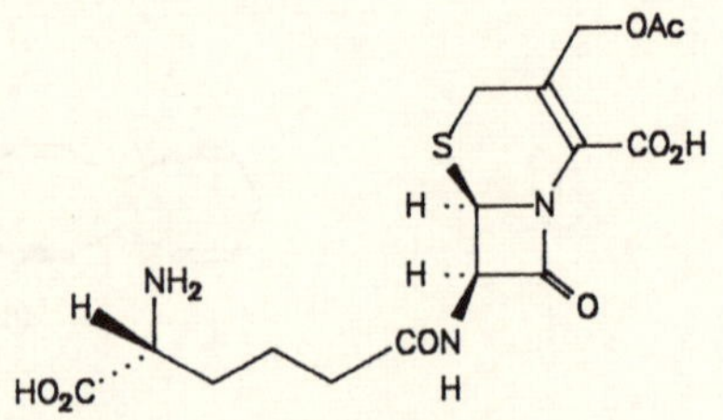

(8·28)

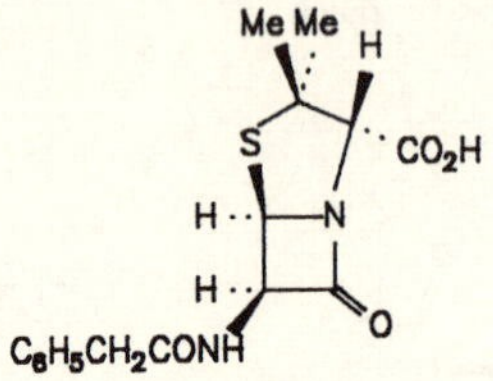

(8·29)

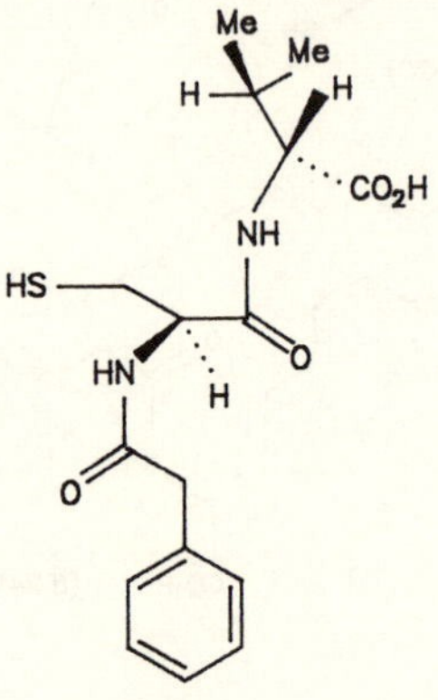

(8·30)

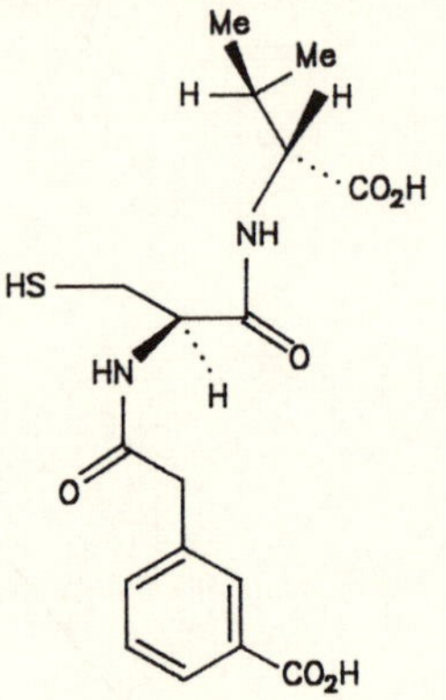

(8·31)

The next step in the biosynthesis of penicillin N involves the oxidative cyclisation of the ACV tripeptide by the remarkable enzyme isopenicillin N synthase (IPNS), in the presence of Fe^{2+} ions, ascorbate and oxygen. IPNS has been subjected to intensive study in order to define its specificity so that new penicillins could be obtained. The unusual reaction producing bicyclic products also prompted an extensive programme of research on its mechanism of catalysis. Variation of the *N*-terminal amino acid of the tripeptide indicated that, for maximum rates, there should be a chain of six carbon atoms terminating in a carboxy group. α-Amino-adipic acid fulfils these requirements. The importance of the carboxy group became evident from a comparison of the rate of formation of penicillin G (**8.29**) from phenylacetylcysteinylvaline (**8.30**) and the much faster rate of reaction of 3′-carboxyphenylacetylcysteinylvaline (**8.31**). The position of the carboxy group in the latter substrate corresponds approximately to that of the ACV substrate. Clearly, the carboxy group plays a positive role in the formation of a penicillin as catalysed by IPNS. Some micro-organisms contain an epimerase that converts isopenicillin N formed by the action of IPNS on ACV into penicillin N. The chiral carbon atom of the δ-(α-aminoadipoyl) moiety is the site of inversion.

IPNS contains Fe^{2+} ($M_r = 32\,000$) and causes the loss of four hydrogen atoms

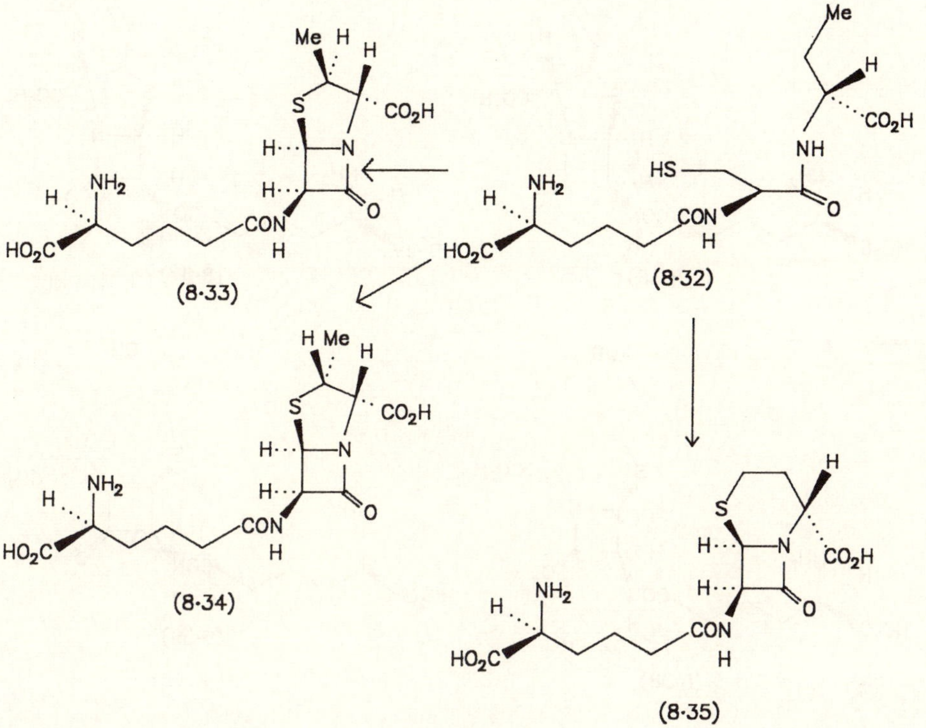

(8·33)

(8·32)

(8·34)

(8·35)

from the peptide substrate; these reduce dioxygen to water. Concomitantly, two new bonds, C—S and C—N, are formed to give the fused bicyclic product comprising a β-lactam and a thiazolidine ring. The specificity of IPNS has been studied extensively using a series of synthetic tripeptides. As indicated above, some variation of the N-terminal residue is permissible, although sometimes at the expense of a diminished rate V_{max}. The central residue must have a thiol group, but some hydrogen atoms other than the 3-(pro-3S)-hydrogen atom can be replaced by a methyl group. Possibilities for producing new penicillins on the basis of structural changes to the central residue are consequently quite limited. Considerably more changes can be introduced in the C-terminal residue of the tripeptide, although a penicillin containing a fused thiazolidine ring may not be formed or may be formed together with a cephalosporin containing a fused 1,3-thiazine ring. Even larger rings are known. For example, the tripeptide (8.32) containing C-terminal α-aminobutyrate gave three products (8.33, 8.34 and 8.35) with IPNS. Again, when 2-allylglycine was at the C-terminal, five products (8.36, 8.37, 8.38, 8.39 and 8.40) were formed. In two cases, an extra oxygen atom was incorporated as a hydroxy group and this stems from the dioxygen substrate. Baldwin and Abraham (1988) have suggested that the duality of the mechanism, comprising *either* oxidative addition of sulphur and of hydroxy across a double bond *or* dehydrogenation of C—H bonds with formation

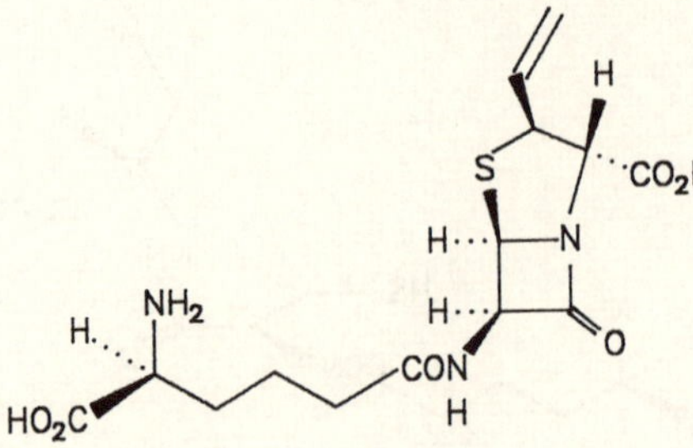

(8·36)

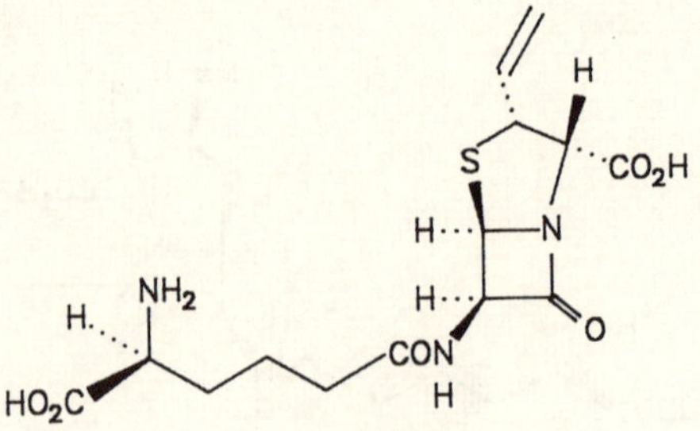

(8·37)

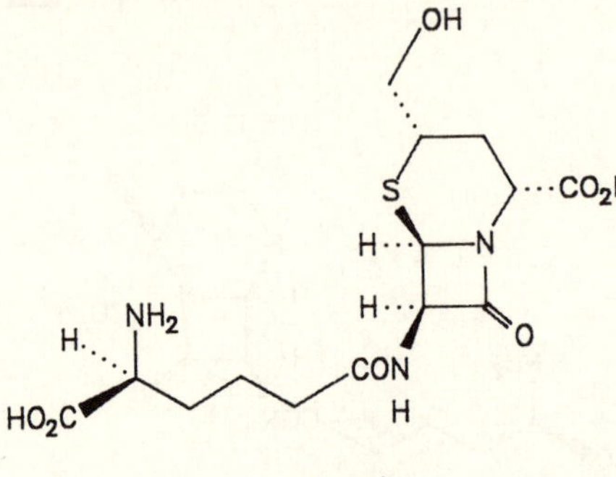

(8·38)

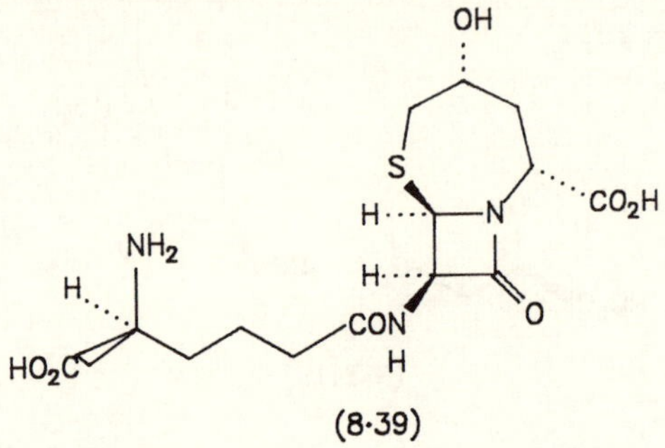

(8·39)

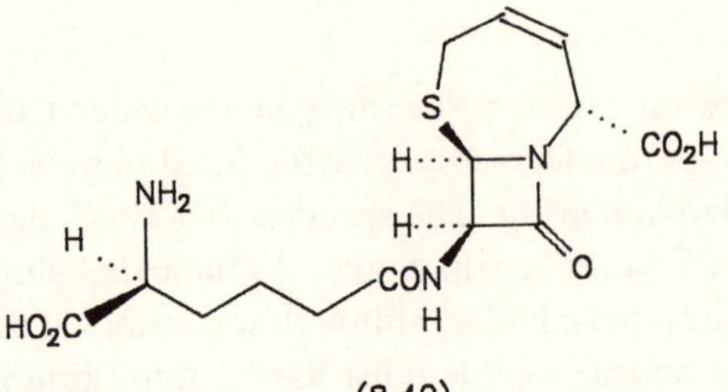

(8·40)

of C—S bonds, involves the formation of a carbon radical. It has further been proposed that a ferryl(IV) species such as Enz=Fe=O is involved at the active site of IPNS. The Enz=Fe=O moiety is believed to be produced from an initial Fe·O_2 complex by a two-electron reduction coupled to the initial formation of the β-lactam ring (Scheme 8.10).

Before it had been discovered that many penicillins could be made from appropriate tripeptides using IPNS, a semi-synthetic method was used to convert penicillin G (8.29) into 6-aminopenicillanic acid using a bacterial acylase followed by acylation of the free amino group. Examples of pharmaceutically important penicillins produced by this route include methicillin (8.41), ampicillin (8.42) and amoxycillin (8.43). There is a more important method of enzymically degrading penicillins than

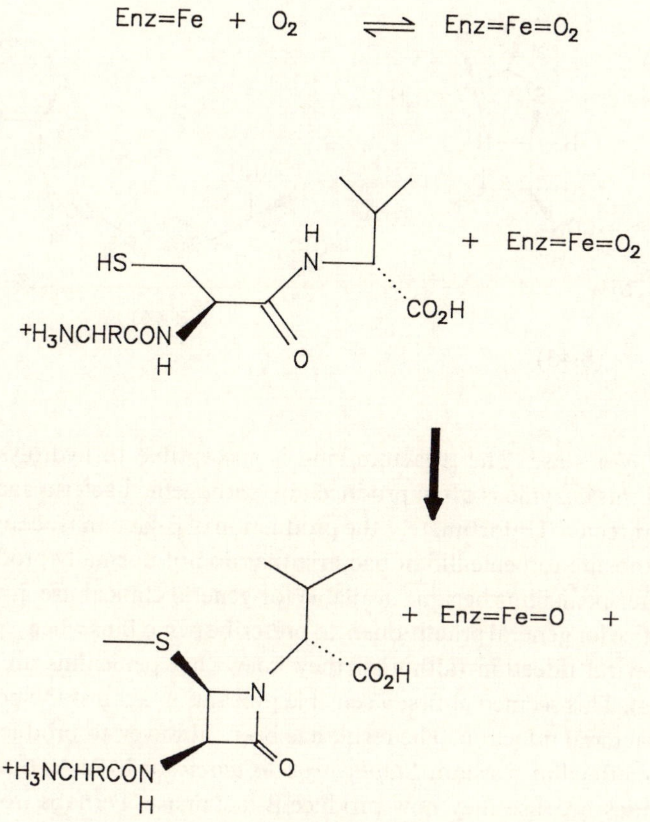

$$\text{Enz=Fe} \ + \ O_2 \ \rightleftharpoons \ \text{Enz=Fe=O}_2$$

Scheme 8.10.

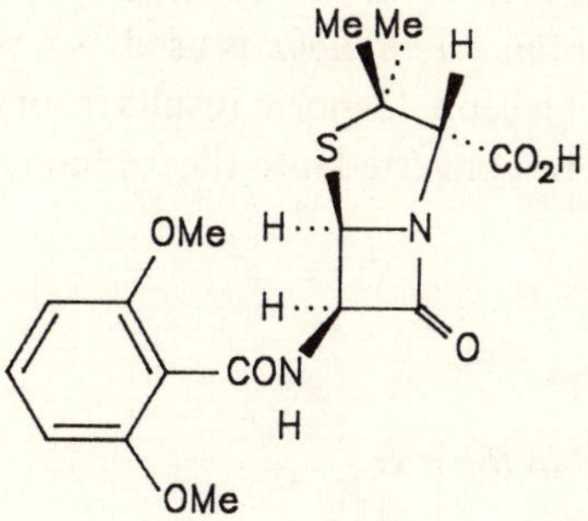

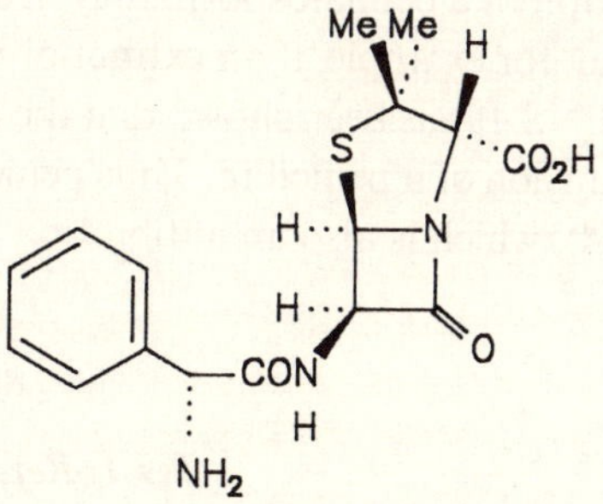

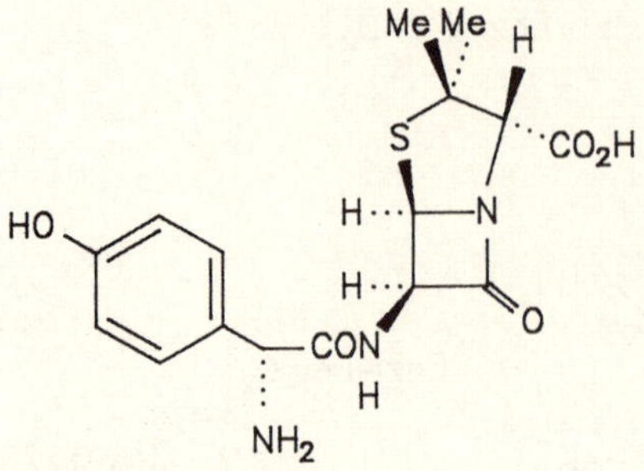

(8·43)

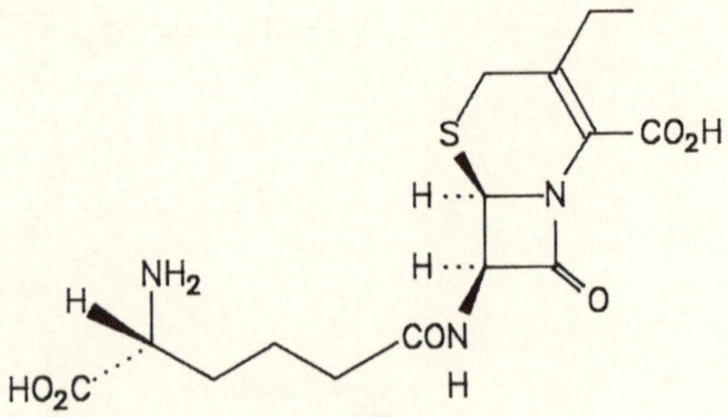

(8·44)

by use of the *N*-acylase. The β-lactam ring is susceptible to hydrolysis by a β-lactamase and this enzyme is often produced by pathogenic bacteria against which penicillins are directed. Unfortunately, the production of β-lactamase can be induced by repeated exposure to penicillin of bacteria that do not normally produce it. For many years after penicillins became available for general clinical use, it was almost standard practice for general practitioners to prescribe penicillins when a patient presented with a viral infection (although they knew that penicillins are ineffective against viruses). This seemed at first a sensible precaution against the possibility of a secondary bacterial infection. The result has been, however, to produce strains of bacteria (e.g. methicillin-resistant *Staphylococcus aureus* or MRSA) that are resistant to antibiotics because they now produce β-lactamase. Perhaps ironically, the most likely place to be infected by such a dangerous organism is in a hospital. Patients are not quite in the same danger from bacterial infections as they were before the Second World War, but new types of antibiotics are urgently needed.

Strictly speaking, once IPNS has operated on a tripeptide substrate, any further changes are outside the domain of peptide chemistry and biochemistry and hence of this book. Nevertheless, for the sake of completeness the conversion of penicillins into cephalosporins is briefly mentioned. It has been mentioned above that the production of a cephalosporin may accompany the formation of a penicillin. This can occur, for example, if an extract of *Cephalosporium acremonium* is used as a source of IPNS. It has been shown that the formation of a cephalosporin results from a ring expansion of a penicillin. Thus penicillin N can be converted into the cephalosporin (**8.44**), which is also an antibiotic.

8.9 References

8.9.1 References cited in the text

Arnstein, H. R. V. and Clubb, M. E. (1957) *Biochem. J.*, **65**, 618.
Arnstein, H. R. V. and Grant, P. T. (1954) *Biochem. J.*, **57**, 360.

Baldwin, J. E. and Abraham, Sir Edward (1988) *Nat. Prod. Rep.*, **5**, 129.
Bradbury, A. F., Finnie, M. D. A. and Smyth, D. G. (1982) *Nature*, **298**, 686.
Kerwin, J. F., Lancaster, J. R. and Feldman, P. L. (1995) *J. Med. Chem.*, **38**, 4343.

8.9.2 References for background reading

Brennan, J. (1986) *Amino Acids and Peptides*, Vol. 17, Chap. 5, Royal Society of Chemistry, London (β-lactam drugs).
Frydrych, C. H. (1991) *Amino Acids and Peptides*, Vol. 22, Chap. 5, Royal Society of Chemistry, London (β-lactam drugs).
Frydrych, C. H. (1992) *Amino Acids and Peptides*, Vol. 23, Chap. 5, Royal Society of Chemistry, London (β-lactam drugs).
Frydrych, C. H. (1993) *Amino Acids and Peptides*, Vol. 24, Chap. 5, Royal Society of Chemistry, London (β-lactam drugs).
Schofield, C. J. and Westwood, N. J. (1995) *Amino Acids, Peptides and Proteins*, Vol. 26, Chap. 6, Royal Society of Chemistry, London (β-lactam drugs).
Stachulski, A. V. (1989) *Amino Acids and Peptides*, Vol. 20, Chap. 5, Royal Society of Chemistry, London (β-lactam drugs).
Stachulski, A. V. (1990) *Amino Acids and Peptides*, Vol. 21, Chap. 5, Royal Society of Chemistry, London (β-lactam drugs).
Voet, D. and Voet, J. G. (1995) *Biochemistry*, 2nd edition, Chaps. 24, 30 and 34. (Amino acid metabolism, protein biosynthesis, blood clotting, peptide hormones and neurotransmitters).

9

Some aspects of
amino-acid and peptide
drug design

9.1 Amino-acid antimetabolites

Amino acids are not suitable in most cases as a basis for developing antibiotics, since they occur in all naturally occurring proteins. Consequently, any attempt to deprive pathological micro-organisms of coded amino acids would cause serious damage to the human host. An exception to this general rule is found with 4-aminobenzoic acid, which is not a coded amino acid but which is present in folic acid (**9.1**). Humans do not require free 4-aminobenzoic acid because they cannot synthesise folic acid. Folic acid is obtained from dietary sources and from the biosynthetic activity of intestinal bacteria. Since bacteria synthesise folic acid from 4-amino-benzoic acid, an antimetabolite of this offers a possible weapon against attack by pathological micro-organisms. 4-Aminobenzenesulphonic acid structurally resembles 4-aminobenzoic acid closely (they are mutually isosteric) and inhibits the synthesis of folic acid by pathogenic organisms, not by competitive inhibition but rather by behaving as an alternative substrate for the enzyme dihydropteroate synthetase, which catalyses the reaction between 2-amino-4-hydroxy-6-hydroxymethyl-7,8-dihydropterin pyrophosphate and 4-aminobenzoic acid. The products of the alternative reaction involving sulphonamides are the expected analogues of dihydropteroate. Crucial to the efficacy of sulphonamides is the failure of folic acid to enter the bacterial cell. All the folic acid required by the bacteria must be synthesised inside the cell. In contrast, sulphonamides like 4-aminobenzoic acid readily enter the bacterial cell. Derivatives of 4-aminobenzenesulphonic acid are a classical example of drug design based on depriving pathogenic organisms of an essential growth factor. By varying the structure of these derivatives, it is possible to control the efficiency of absorption from the gut, the rate of metabolism of the drug and the rate of excretion of it and its metabolities. For example, a disease such as dysentery requires a drug that is poorly absorbed from the gut whereas a bacterial infection of the urinary tract requires a drug that is rapidly absorbed and steadily excreted via

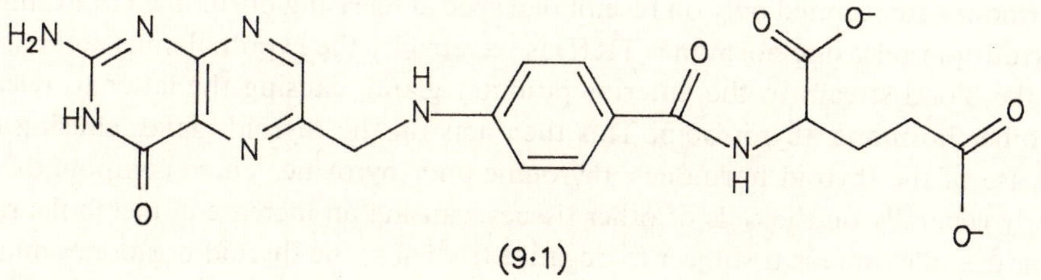

(9·1)

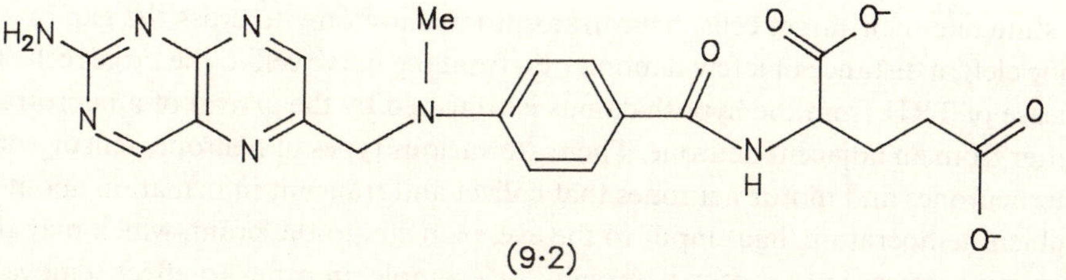

(9·2)

the kidneys. The cytotoxic drug methotrexate (**9.2**) is an analogue of folic acid that has long been used for treating acute leukaemia in children, Burkitt's lymphoma and a rare type of malignancy, choriocarcinoma, that occurs in relation to pregnancy. Unfortunately, in most cases, the side effects on normal tissues can be severe.

Pantothenic acid (pantoyl-β-alanine) is another amino-acid derivative that is required by bacteria and pantoyltaurine is an antimetabolite. Unfortunately, pantothenic acid is also required by humans so the possibility of the development of a drug from this growth factor is severely limited.

9.2 Fundamental aspects of peptide drug design

The human body produces a large number of peptides, many of which have been classified as *hormones* (Greek ὁρμάω, meaning 'I excite'). Examples of peptide hormones include insulin, glucagon, gastrin and cholecystokinin. Other types of molecules such as steroids and amines also behave as hormones, but these are not dealt with in this book. Hormones are produced by specialised cells and very low concentrations of hormones can carry biological instructions to other cells. It is common to describe a hormone as a *first messenger*, since, on arrival at the target cell, it is bound by a specific receptor. It is then internalised into the target cell, whereupon a second messenger completes the delivery of the biological message. Sometimes, the secreted hormone is received by a cell in the vicinity of the cell which secreted the hormone, but more commonly, the hormone is carried in the blood stream to cells that are remote from the hormone source. For example, in child-birth or parturition, oxytocin is released by the posterior pituitary gland and travels to the uterus, causing contractions that eventually expel the foetus. Some peptide hormones are secreted only on receipt of a special releasing hormone. For example, thyrotropin-releasing hormone (TRH) is secreted by the hypothalamus and passes in the blood stream to the anterior pituitary gland, causing the latter to release another hormone, thyrotropin. This then acts on the thyroid gland, causing the release of the thyroid hormones, thyronine and thyroxine. These compounds act fairly generally on the cells of other tissues, causing an increase in metabolic rate. This cascade process is subject to negative feedback; the thyroid hormones inhibit the release of thyrotropin by the anterior pituitary.

In contrast to some hormones that have to travel considerable distances in order to stimulate their target cells, neurotransmitters have only to cross the gap or synaptic cleft, a distance of a few nanometres, from the nerve cell to the target cell. The release of TRH from the hypothalamus is triggered by the arrival of a neurotransmitter from an adjacent neurone. There are various types of neurone, sensory ones, interneurones and motor neurones that collect and transmit information about the ambient temperature, light input to the eye, pain etc. to the brain, which may then transmit a message to motor neurones, for example, in order to effect removal of one's finger from a hot object.

A third type of molecular messenger comprises peptides that are cell-growth factors. These molecules do not affect all cells since this would clearly be harmful. Growth factors tend to act on cells that turn over rapidly or are prone to damage by wounding. For example, nerve-growth factor (NGF) promotes growth but not division of nerve cells. Platelet-derived growth factor (PDGF) stimulates both growth and division of cells in connective tissue such as fibroblasts and smooth muscle cells. It assists in the repair of damaged blood vessels. Epidermal growth factor (EGF), like PDGF, stimulates cell division. Both have their own specific receptors on cells and both stimulate the phosphorylation of certain hydroxy groups in proteins. The advent of malignancy almost certainly accompanies a failure of control of this system.

9.3 The need for peptide-based drugs

The foregoing section indicates some reasons why peptide-based drugs are needed. If a peptide hormone is not produced in sufficient quantities or is defective in structure, then a replacement is required. Peptides, especially very small molecules, have a very short half life in the body. The reason for this is the ubiquitous occurrence of proteolytic enzymes that effect hydrolysis of peptides to the constituent amino acids. Although longer peptides, especially those with structural features such as disulphide bonds, survive longer *in vivo*, they are more likely to stimulate the body's immune system to produce antibodies and effect removal of the peptides. This is particularly likely to occur with molecules that differ structurally from the naturally occurring hormones. Thus, treatment of juvenile-onset diabetes mellitus with insulins from animal sources can occasionally stimulate the patient's immune system

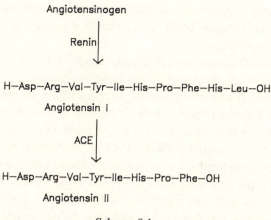

Angiotensinogen

Renin

H—Asp—Arg—Val—Tyr—Ile—His—Pro—Phe—His—Leu—OH

Angiotensin I

ACE

H—Asp—Arg—Val—Tyr—Ile—His—Pro—Phe—OH

Angiotensin II

Scheme 9.1.

to react to the foreign protein even though the latter has high activity *in vivo* in man. This phenomenon has prompted the chemical synthesis or semi-synthesis of human insulin (see Section 7.11). The appearance of juvenile-onset diabetes mellitus is due to an autoimmune condition in which the body attacks the β-cells of its own pancreas. Conversely, late-onset diabetes mellitus, which tends to affect obese people after about 65 years of age, is probably due to a dearth of insulin receptors. Although the blood levels of insulin are often high, the hormone cannot pass into cells to perform its biochemical role.

Low levels of peptide hormones may result from too rapid destruction by proteolytic enzymes. High levels of proteinases lead to the destruction of larger molecules than hormones. In pancreatitis, there is a large overproduction of chymotrypsin and trypsin so that proteolysis of essential proteins is out of control. Again, overproduction of zinc-containing proteinases can facilitate the invasion of adjacent tissues by a malignant tumour. Conversely and perversely, many peptide hormones are produced by proteolysis of biologically inactive precursors. High levels of the proteinase that generates a hormone from an inactive precursor will produce elevated levels of the active peptide. A well-known example is shown in Scheme 9.1; a physiologically inactive protein, angiotensinogen, is converted into angiotensin I by the proteinase renin, which is produced by the kidney. (Be careful to distinguish between renin and rennin which is used in the manufacture of yoghourt and cheese.) Angiotensin I is then converted into the vasopressive angiotensin II by the angiotensin-converting enzyme (ACE). The life-threatening hypertension produced by overproduction of angiotensin II has led to a massive programme of molecular design and synthesis of inhibitors both of renin and of the ACE. Likewise, there has been an extensive search for inhibitors of the metalloproteinases in order to prolong the life expectancy of patients with cancer. After smoking for a number of years, there is a considerable risk that emphysema will be present in the lungs. This causes the liberation of elastase from leucocytes and this enzyme, as its name suggests breaks down elastin, a protein that confers the elasticity to normal lung tissue.

The ubiquitous occurrence of proteinases is accompanied by a similar distribution of fairly specific proteins that inhibit these enzymes. In normal health, there is a delicate balance between the levels of enzymes and their macromolecular inhibitors. This balance is particularly important in the blood-clotting clot-lysis scheme. Since the total volume of blood in the adult human body is only about 5 l, a massive response is required in the event of an injury that results in a rapid haemorrhage. Some positive feedback is present in the blood-clotting mechanism in order to achieve this rapid response, but clearly this must be sensitively controlled by endogenous inhibitors if a thrombosis is not to occur.

Some of the general problems associated with the design of peptide-based drugs can now be appreciated. We shall begin by considering the design of inhibitors of proteinases.

9.4 The mechanism of action of proteinases and design of inhibitors

There are four main types of proteinases: (a) serine proteinases that contain a serine residue at the active site, the hydroxy group of which has enhanced nucleophilicity, and the substrate acylates this residue with simultaneous liberation of the amino component of the peptide bond that is cleaved by the proteinase; (b) cysteine proteinases that contain a cysteine residue at the active centre and the thiol group undergoes intermediate formation of an S-acyl intermediate similar to principle to the mechanism undergone by serine proteinases; (c) aspartate proteinases that contain aspartic acid residues at the active site; and (d) metalloproteinases that contain a zinc cation coordinated to the side-chains of amino acids such as aspartic acid and histidine.

The acylated enzymes of serine and cysteine proteinases and their hydrolysis products are produced with the formation of transition states in which the carbonyl carbon atom of the acyl group is believed to adopt a sp^3 structure. There are several types of potential inhibitors for these enzymes. Peptides that simulate the structure of a good substrate in amino-acid sequence are likely to bind well to the active site of the enzyme. In addition, if the scissile peptide bond of a good substrate is replaced by a similar structure that is resistant to hydrolysis by the enzyme, a good competitive inhibitor will be the result. It is possible, however, that one or more peptide bonds in the remainder of the molecule will be hydrolysed by other proteinases and the strength of binding to the target enzyme will be impaired. A range of structural strategies to confer stability of a peptide or pseudo-peptide against enzymic hydrolysis is possible. Almost all proteolytic enzymes are specific for the hydrolysis of peptide bonds derived from L-amino acids. Introduction of D-amino acids at suitable points in the sequence will often confer stability with respect to enzymic

(9.3)

RCO—Val—Val—Sta—Ala—Sta—OH

$R = Me_2CH(CH_2)_n-, Me(CH_2)_n-$ (n = 0,1—20)

(9.4)

hydrolysis, perhaps with an acceptable lowering of the binding affinity. Again, the introduction of substituents into the component amino acids of the peptide inhibitor at suitable places can often afford protection. The presence of a methyl substituent on the α-carbon atom or the use of an *N*-methyl amino acid for peptide synthesis can confer complete stability at the expense of a somewhat more difficult coupling step in the synthesis of the peptide. In the case of chymotrypsin, the presence of methyl groups in the 2,6 positions of the aromatic ring of phenylalanine of a synthetic peptide renders the peptide bond involving that residue stable with respect to hydrolysis.

The recognition that serine and cysteine proteinases catalyse hydrolytic reactions through a transition state intermediate that has a sp³ structure has led to the design of extremely potent enzyme inhibitors. It was proposed that the efficient catalysis by proteinases depended on their affinity for binding the transition state in preference to the ground state (Pauling, 1946; Wolfenden, 1972). A good example of a proteinase inhibitor derived from statine (**9.3**) with a structure that resembles the putative transition state is pepstatin (**9.4**). This occurs naturally in some micro-organisms. It inhibits aspartate proteinases and its structure should be compared with that of the proposed transition state during pepsin-catalysed hydrolysis (Figure 9.1). Pepstatin is active in such low concentrations that it has proved an invaluable lead in the search for inhibitors of renin. Several syntheses of statine have been developed and modification of the structure in order to optimise its inhibitory activity then involves only straightforward peptide synthesis.

A wide range of compounds containing sp³ carbon in place of sp² carbon in the scissile bond of a substrate has been produced in pharmaceutical chemistry laboratories. An obvious structure is obtained by replacing the scissile —CONH— peptide bond by —CH₂NH— and a route for this is outline in Scheme 9.2 (Sasaki *et al.*, 1987; Rodriguez *et al.*, 1987). As expected, these pseudo-peptides are potent inhibitors of several proteinases depending on the degree of resemblance between the rest of the structure and that of a good substrate. A potential disadvantage of these pseudo-peptides concerns the basicity of the secondary amino group.

Some other analogues of putative transition states contain moieties such as

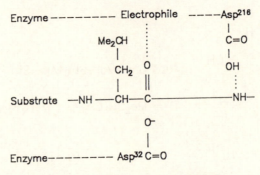

Figure 9.1. Pepsin-catalysed hydrolysis.

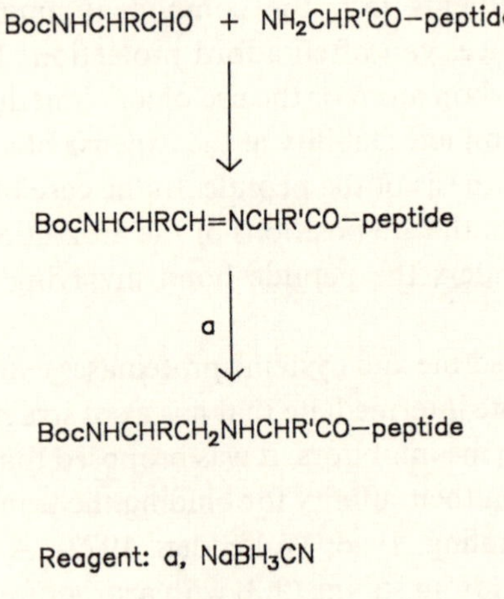

BocNHCHRCHO + NH$_2$CHR'CO—peptide

BocNHCHRCH=NCHR'CO—peptide

a

BocNHCHRCH$_2$NHCHR'CO—peptide

Reagent: a, NaBH$_3$CN

Scheme 9.2.

ψ[CH$_2$O], ψ[CH$_2$CH$_2$] and ψ[CH$_2$S]. Pseudo-peptides generated from α-aminoboronic and α-amino phosphonic acids also simulate the structure of transition states.

In designing potential inhibitors of metallo-enzymes, it is common to have a lead (not Pb!) compound with a structure resembling that of at least part of the substrate and, in addition, containing a group or groups that could function as a good ligand for the metal in the enzyme. As indicated in Scheme 9.1, the production of the hypertensive peptide angiotensin II from angiotensin I requires the ACE, which contains Zn as part of its catalytic centre. Potential Zn ligands include carboxy, thiol and imidazole groups. One of the simplest and most potent inhibitors for the ACE is D-3-mercapto-2-methylpropanoyl-L-proline ('Captopril') (IC$_{50}$ 23 nM). It is orally active.

Some proteinases, especially those with a Ser or Cys residue at the active site, as indicated above, catalyse the hydrolysis of peptide bonds in two steps. In the first stage, the active site Ser residue is acylated by the peptide moiety that terminates in

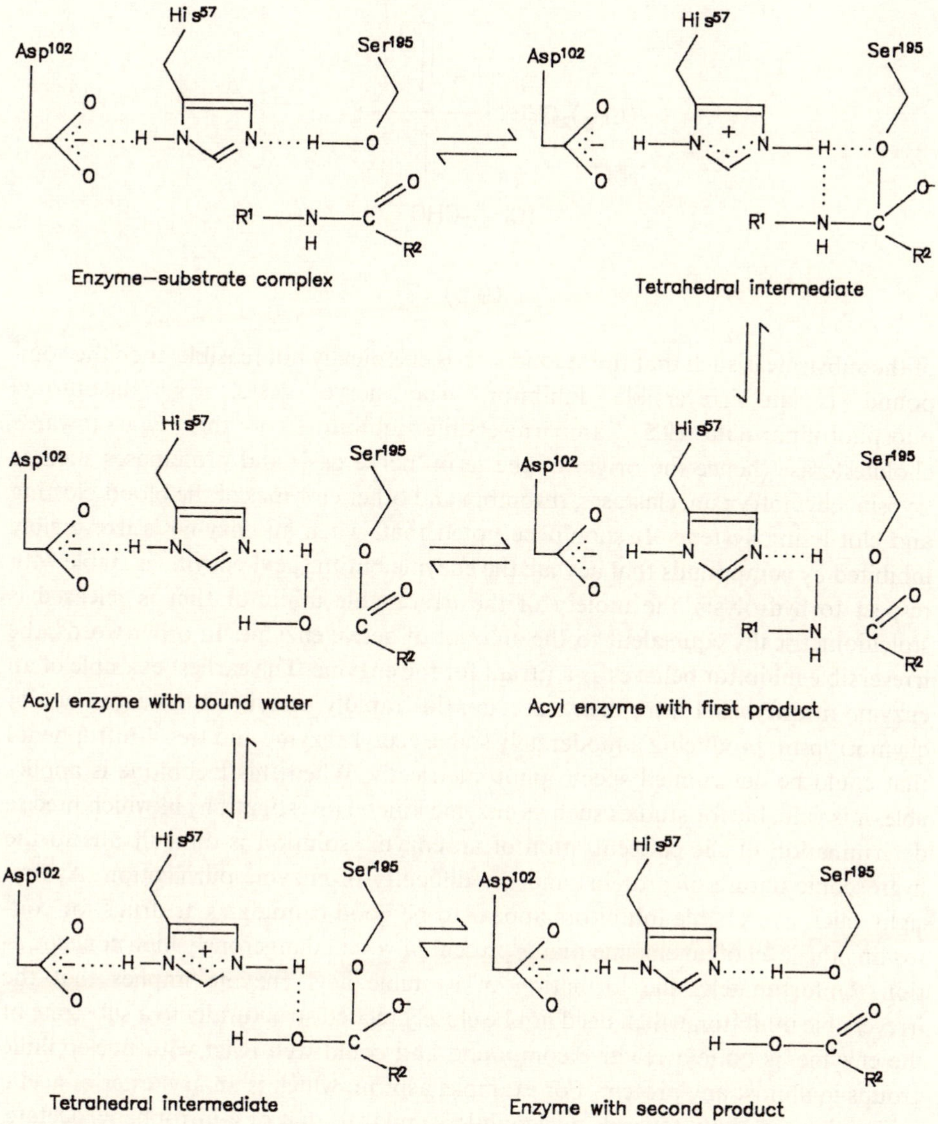

Scheme 9.3.

the carbonyl group of the scissile bond. This stage proceeds through a transition state in which the carbonyl group is transiently converted into an sp^3 carbon atom linked to a hydroxy group. The other moiety of the substrate, which now contains a free $-NH_3^+$ group derived from the $-NH-$ of the scissile bond, is concurrently liberated as the first product. In the second stage, the acylated enzyme is hydrolysed back to the free enzyme, presumably through a second transition state, and the second product containing a free $-COO^-$ is liberated (Scheme 9.3). If the structure

207

$$(CH_3)_2CHO \longrightarrow \overset{\displaystyle O}{\underset{\displaystyle (CH_3)_2CHO}{\overset{\displaystyle \|}{P}}} \longrightarrow F$$

(9.5)

of the substrate is such that the second step is chemically not feasible, then the compound is an irreversible inhibitor. The nerve gases, e.g. diisopropyl phosphorofluoridate (9.5), are irreversible inhibitors of this type towards cholinesterase (hence the origin of the term 'nerve gas') and proteinases such as trypsin, chymotrypsin, elastases, thrombin and other enzymes of the blood-clotting and clot-lysing systems. It should be noted that, when an enzyme is irreversibly inhibited by compounds that acylate the enzyme but the acyl-enzyme is stable with respect to hydrolysis, the moiety of the irreversible inhibitor that is released is stoichiometrically equivalent to the amount of active enzyme. In other words, the irreversible inhibitor behaves as a titrant for the enzyme. The earliest example of an enzyme titrant was 4-nitrophenyl acetate; this rapidly acetylated the active site of chymotrypsin, producing a moderately stable acetyl enzyme and free 4-nitrophenol that could be determined spectrophotometrically. When this technique is applicable, it is valuable for studies such as enzyme kinetic investigations in which precise determination of the concentration of an enzyme solution is difficult due to the hygroscopic nature of proteins and the difficulty of enzyme purification. At first sight, such irreversible inhibitors appear to be good candidates as drugs for controlling the level of an enzyme that is present *in vivo* in dangerously high concentration. Unfortunately, the formation of a stable acyl enzyme implies that the irreversible inhibitor, which need not be closely related structurally to a substrate of the enzyme, is quite a reactive compound and could well react with nucleophilic groups in almost any protein. For example, aspirin, which is an aryl ester of acetic acid with a reactivity towards nucleophiles similar to that of 4-nitrophenyl acetate, not only relieves pain by acetylating Ser[530] of prostaglandin H_2 synthase and thus inhibiting the enzyme but also inhibits platelet aggregation, which is an essential step in the unrelated process of blood clotting. This latter property is harnessed for prophylactic treatment of patients who might be prone to having a stroke. Unfortunately, it rules out the use of aspirin for treatment of the common cold or a headache for patients with a history of peptic or duodenal ulceration since impairment of the clotting process by aspirin might precipitate a serious haemorrhage.

Some ingenious chemistry has been involved in the design of the next type of enzyme inhibitor. If a compound contains a bond that can be cleaved by the enzyme,

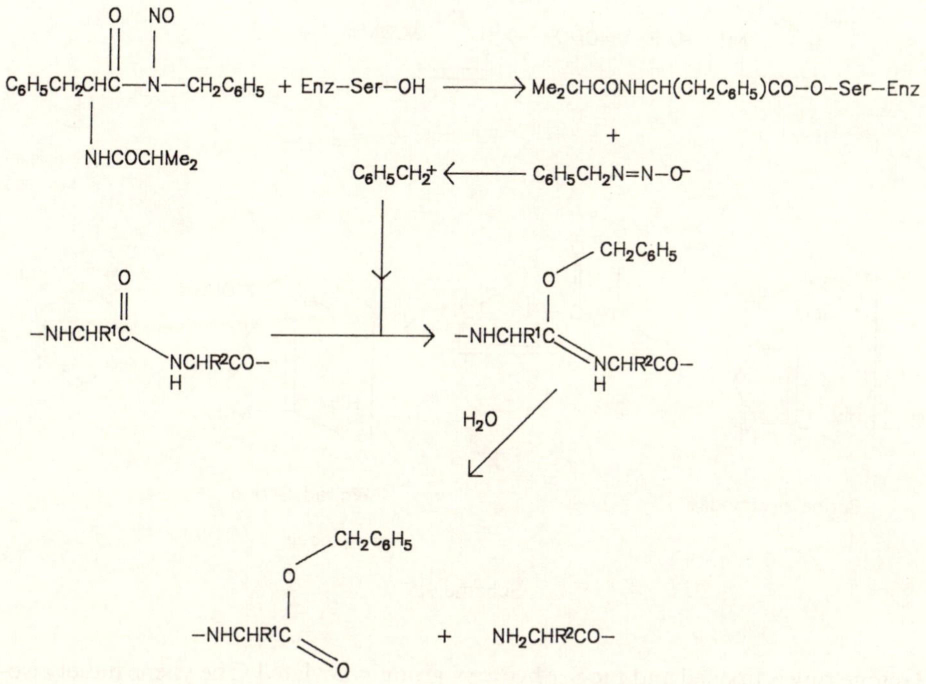

Scheme 9.4.

thereby generating a compound that is capable of reacting covalently with a group near to the active centre and thus leading to irreversible inhibition, then the compound is described as a k_{cat} or suicide inhibitor. In other words, the compound only becomes an irreversible inhibitor when it has been used as a substrate by the enzyme. The bond initially cleaved by a proteinase is either an ester or an amide bond. The hydroxy group in the side-chain of the essential Ser residue commonly remains acylated so that the residual structure of the suicide inhibitor is attached to the enzyme at two points. That this is not always the case is illustrated by the inhibition of chymotrypsin by Me_2CHCO—Phe—$N(N=O)CH_2C_6H_5$ (Scheme 9.4) (Donadio *et al.*, 1985). Following enzymic cleavage of the amide bond, a benzyl cation that can alkylate various groups is liberated. Because the benzyl cation is liberated into solution, a particular enzyme molecule may cleave a number of suicide inhibitor molecules before it becomes the site of attack by the benzyl cation. A somewhat similar state of affairs exists in the inhibition of pancreatic elastase with imidazole *N*-carboxyamides (Scheme 9.5) (Groutas *et al.*, 1980). The enzyme cleaves the ImCO—N bond and an alkyl isocyanate is generated. This then irreversibly carbamoylates the hydroxy group of the essential Ser residue. An example of a suicide inhibitor that becomes covalently attached at two sites in the enzyme is seen in Scheme 9.6. Leukocyte elastase is inhibited by ynenol lactones (Tam *et al.*, 1984). First of all the

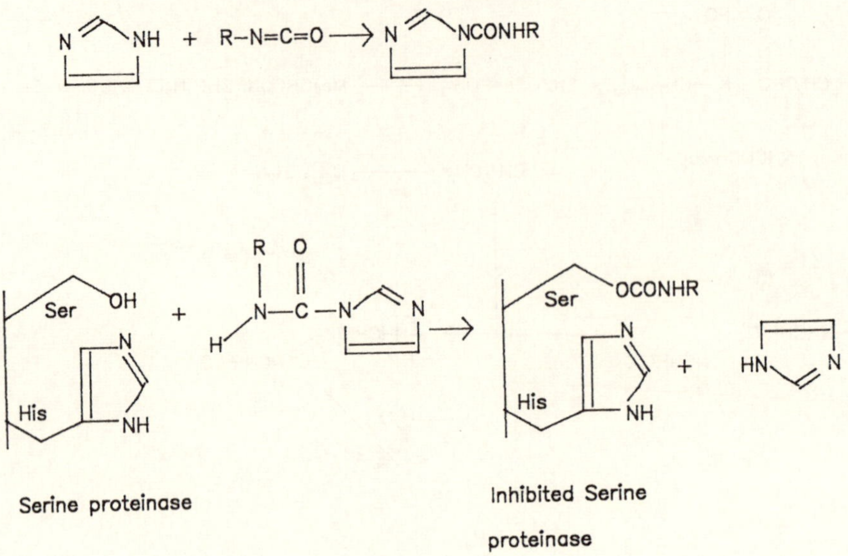

Scheme 9.5.

lactone ring is opened and the Ser hydroxy group is acylated. The ynene moiety iso-merises to an allenone that then captures a nucleophilic group adjacent to the active site (Enz—Nu).

9.5 Some biologically active analogues of peptide hormones

In contrast to the previous section, in which analogues of transition states were the preferred structures for potential enzyme inhibitors, potentially useful analogues of peptide hormones are likely to contain pseudo-peptide bonds that compare to the ground state of conventional peptide bonds. For example, if one is attempting to design an analogue of a peptide hormone that is rapidly degraded *in vivo*, then replacement of the most hydrolytically sensitive peptide bond by a closely analogous group may confer protection against enzymic attack without interfering seriously with the binding of the analogue to a cellular receptor. The analogue may then display the activity of the original peptide hormone and be longer acting. On the other hand, the analogue may bind to the receptor because of its structural resemblance to the natural peptide, but fail to be internalised by the target cell. It would then behave as an antagonist by interfering with the capture of the natural peptide (Hardie, 1991).

Thionopeptides, with the —CSNH— group replacing one or more peptide bonds, closely resemble the related peptides. The —CSNH— group usually has *trans* sub-stituents; the major differences are the length of the C—N bond and the size of the sulphur atom. Thionopeptides are resistant to hydrolysis by proteinases. Despite

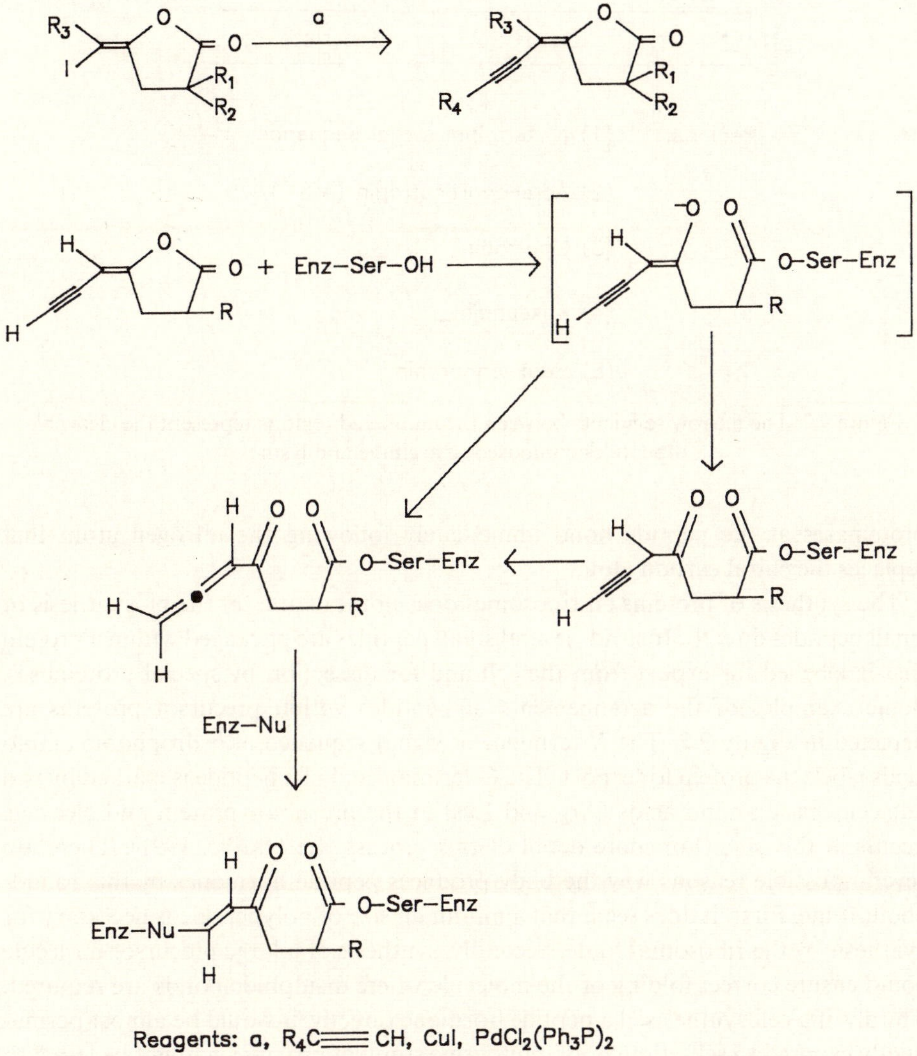

Reagents: a, R₄C≡≡≡CH, CuI, PdCl₂(Ph₃P)₂

Scheme 9.6.

their apparent attraction as potential drugs, they have not received the attention afforded to other ground-state analogues of biologically active peptides.

Retropeptides contain the —NHCO— group and when the adjacent amino acids have the D configuration the structural resemblance to the related L peptide is quite close. Moreover, such retro-inverso peptides are stable to hydrolysis by proteinases.

A third type of peptide analogue that has been studied widely is the azapeptide, in which the chiral carbon atom of an amino-acid residue is replaced by nitrogen. As with thionopeptides and retropeptides, azapeptides are resistant to the action of

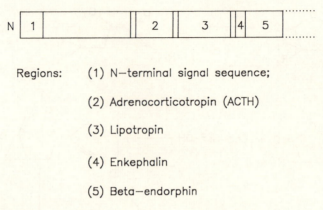

Regions: (1) N—terminal signal sequence;

(2) Adrenocorticotropin (ACTH)

(3) Lipotropin

(4) Enkephalin

(5) Beta—endorphin

Figure 9.2. The narrow segments between the numbered regions represent the cleavable dipeptides composed of arginine and lysine.

proteinases at the peptide bond immediately following the nitrogen atom that replaces the chiral carbon atom.

The synthesis of proteins on ribosomes does not function for the biosynthesis of small peptides directly. Instead, several small peptides are packaged within a protein that is labelled for export from the cell and for dissection by special proteinases. Some examples of the arrangements of peptides within precursor proteins are depicted in Figure 9.2. The *N*-terminal or signal sequence of hydrophobic amino acids labels the protein for export. The *C*-terminal end of a peptide is marked by two adjacent basic amino acids (Arg and Lys) in the precursor protein and cleavage occurs at this site. (For more detail of this process, see Hardie, 1991). There are several possible reasons why the body produces peptide hormones by this round-about route. First, it does seem that a minimum size of polypeptide is necessary for synthesis by the ribosomal route. Secondly, synthesis of a large precursor molecule could ensure correct folding of the molecule where disulphide bonds are required. Thirdly, if a cell synthesised a peptide hormone directly, it would be almost permanently exposed to self-stimulation (autocrine stimulation), which might be lethal to the cell.

The enkephalins, H—Tyr—Gly—Gly—Phe—X—OH (X=Leu, Met), or so-called opioid peptides because they mimic the action of the opiates, morphine and heroin, have a very short half life in the body because all four peptide bonds are prone to undergoing proteolysis. The Tyr—Gly bond can be hydrolysed by amino-peptidases, the Gly—Gly bond by dipeptidylaminopeptidases, the Gly—Phe bond by enkephalinase and the Phe—Met and Phe—Leu bonds by carboxypeptidases. An enormous number of analogues have been synthesised, especially with the object of producing compounds that exert potent analgaesic action but are free from side effects. Protection of the susceptible bonds by changing the amino-acid sequence is the obvious way to achieve this. The analogue H—Tyr—D—Met—Gly—Phe—

212

Pro—NH$_2$ is resistant to three of the four types of enzyme listed above. There are several receptors for enkephalins, labelled μ, κ, σ and δ and analogues that are selective for particular receptors have been synthesised. For example, H—Tyr—D-Ala—Gly—Phe—Leu—OH is selective for δ receptors whereas H—Tyr—D-Ala—Gly—MePhe—Met(O)—OH is selective for μ receptors.

A few examples of analogues of other peptide hormones will now be given but it must be appreciated that a single paper may describe several dozen new compounds and many thousands are known. Three analogues of angiotensin II are

H—Sar—Arg—Val—Tyr—Val—His—Pro—Ala—OH ('Saralasin')
H—Sar—Arg—Val—Tyr—Ile—His—Pro—D-Phe—OH
H—Sar—Arg—Val—Tyr—Val—His—Pro—Ala(Ph$_2$)—OH.

Note that all have *N*-terminal sarcosine and are therefore resistant to aminopeptidases. The second peptide has *C*-terminal D-Phe whereas the third has *C*-terminal β-diphenyl-alanine. Both are resistant to carboxypeptidases. All three compounds are antagonists of angiotensin II.

Omission of the *C*-terminal Arg residue from the vasodilator bradykinin

H—Arg—Pro—Pro—Gly—Phe—Ser—Pro—Phe—Arg—OH

affords an agonist, i.e. a compound that activates the receptors and potentiates the binding of the natural peptide. When the *C*-terminal Phe residue of the octapeptide agonist is replaced by Leu, the resultant peptide is an antagonist for one (B$_1$) of the two types of receptor for bradykinin. If the Pro^7 residue of bradykinin is replaced by D-Phe, the resultant peptide is an antagonist for the B$_2$ receptor. These examples illustrate how quite small changes in peptide structure can completely change the pharmacological behaviour.

9.6 The production of antibodies and vaccines

Although numerous antibiotics have been isolated from natural sources or synthesised in the laboratory for combatting bacterial infections, nothing like the same degree of success has attended the attempts to overcome attack by viruses. Fortunately, there is an alternative strategy. The body possesses a defence mechanism that is capable of distinguishing between proteins from self and proteins originating from foreign sources. Specialised cells produce antibodies against foreign proteins and these are disposed of by the body. The subject is too large to describe here but a good general text on biochemistry or immunology will give an adequate background. We wish here to consider how one can cause the immunological defence mechanism to respond to a naturally occurring or synthetic peptide or protein by producing antibody proteins (immunoglobulins) so that, in the event of

213

exposure to a virus or bacterium containing that sequence of amino acids, the body will be able to overcome the microbiological attack. This process is known as vaccination or immunisation. It is known that only small sections of a protein are necessary in order to evoke antibody production but this peptide sequence should be attached to or be part of a macromolecule for efficient antibody production. Small peptides are not immunogenic. Frequently, a linear segment of the foreign protein is adequate for stimulating antibody production but sometimes a better reaction is obtained by designing a peptide from amino acids that are juxtaposed on the surface of the protein, although they need not be sequential in the protein.

Peptide sequences that are immunogenic and evoke efficient production of antibodies are known as *epitopes*. It is usual to find that peptides containing about eight amino acids are required. The problem facing the chemist or biochemist aiming to produce a vaccine is to identify and synthesise the most suitable epitopic sequence attached to a macromolecular carrier. There are other problems for the immunologist or clinician such as finding the best method of administration to the patient, coping with any adverse reaction to vaccination and assessing the degree of protection afforded by the vaccination. It should be pointed out that the use of synthetic peptides might not produce a very active vaccine. Sometimes, the use of a killed virus as an immunogen will produce better results. These aspects will not concern us here. One obvious synthetic approach is to begin at the N-terminus of the protein in question and synthesise say octapeptides along the whole protein sequence, shifting the frame sequence by one or two amino acids at a time. Solid-phase synthetic methodology using automatic machines makes this a feasible project even with fairly long proteins. It is sometimes possible to shorten the process by judging which parts of the sequence are likely to lie on the surface of the protein, for these are likely to be the most immunogenic ones. Information about which amino-acid residues are on the surface of a protein can be obtained, for example, by exposure of the protein in question to reagents (e.g. acylating agents, diazonium salts, iodoacetic acid and diazoalkanes) that react covalently with side-chains of amino acids. The sites of chemical modification of the protein can then be identified by the sequencing methods described in Chapter 5.

Potential epitopic sequences can be covalently atttached to the side-chains of proteins in order to enhance their immunogenicity or they can be synthesised by the solid-phase method and left attached to the resin for injection (Goddard *et al.*, 1988). Another idea is to attach multiple copies of a peptide to a support such as (9.6). The S-acetyl groups are removed with NH_2OH and the peptide antigen bearing an N-terminal S-(3-nitropyridine-2-sulphenyl)cysteinyl residue is added. An exchange reaction forms disulphide bonds and liberates 3-nitro-2-thiopyridone (Drijfhout and Bloemhoff, 1991).

Clearly, it is advantageous to use some system of multiple synthesis of peptides in order to minimise the time required to assemble a library of peptides derived from a large protein. It was the need to synthesise many peptides, especially when searching for a lead compound with desirable pharmacological properties, that led to a com-

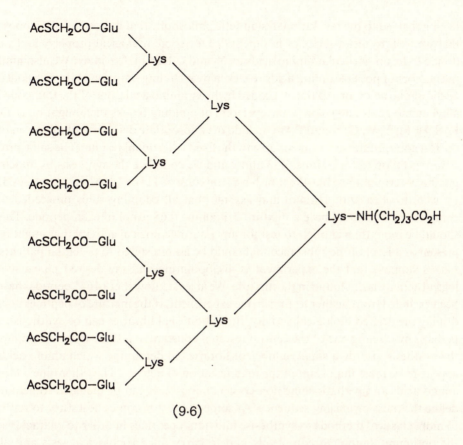

(9·6)

pletely new philosophy in organic chemistry. For the last one and a half centuries, organic chemists have obeyed a kind of holy writ in synthetic studies. Methods have been designed to give the highest possible yield of one compound, which has then been purified by the best techniques available at the time and the compound has been characterised by elementary analysis and spectroscopy. Although there is no alternative to this classical methodology if it is desired to determine a quantitative relationship between a structure and its properties, it is extremely labour intensive. For example, simply to assemble a library of hexapeptides containing only the twenty coded amino acids would involve making 64 000 000 compounds. Assembly and biological testing of a library of compounds greatly accelerates the search for at least a lead compound. It should be noted that this kind of approach is by no means limited to the search for pharmacologically active peptides.

9.7 The combinatorial synthesis of peptides

This topic could have been included in Chapter 7, but has been included here because it was the need to produce large numbers of peptides for pharmacological testing that led to the revolution in synthetic philosophy that has occurred. Suppose that we wish to test a series of hexapeptides for some biological property and we decide to fix on particular amino acids as *N*- and *C*-terminal residues. We also limit the number of possible amino acids to eleven by including only one of the two coded acidic amino acids, one of the two coded hydroxy amino acids, one of the four coded alkyl amino acids and so on. Our repertoire of building blocks then might be L, D, K, S, H, M, Y, W, G, N and P. We include the appropriate derivative of all eleven of the foregoing amino acids and couple to the fixed *C*-terminal residue. The same procedure is followed for residues 4, 3 and 2 and we complete the synthesis by attaching the *N*-terminal residue. We now have a mixture of 11^4 or 14641 hexapeptides. If we were to start with 1 mmol and assume that all coupling steps proceeded to completion, we should have a mixture containing 0.68 μmol of each peptide. This should be more than enough to test for any pharmacological activity that might be present at a level of the substance that could be administered to potential patients. Let us suppose that the experiment is disappointing and the desired pharmacological activity is not found in the mixture. We at least know that 14641 peptides have been excluded from further testing in one experiment. If the mixture of peptides does display the desired biological activity, then additional libraries can be synthesised, perhaps by keeping one of the central residues constant at a time. It should require the synthesis of only a small number of libraries to determine which amino acids appear to be most important in the manifestation of activity. The other nine coded amino acids and perhaps some non-coded amino acids can be included to help to define the most promising sequence. At some stage, it becomes necessary to revert to more classical methods to synthesise individual peptides in order to characterise the optimum compound completely and to carry out toxicological tests and all the other tests on animals and eventually on humans before a new drug comes on to the market. Already, more esoteric variations of the technique are available, such as restricting the synthesis so that each peptide in the library is produced on an individual bead of macromolecular support and even tagging each bead with a different simple compound that can be identified by some simple chemical or spectroscopic test in order to index the library of peptides (Janda, 1994; Nestler *et al.*, 1994).

9.8 The design of pro-drugs based on peptides

A pro-drug is a substance that has no special biological activity *per se* but can be converted into an active drug by enzymic action in the body. Thus, all the initial proteins formed by ribosomal synthesis that contain a peptide hormone structure locked within their amino-acid sequence are analogous to pro-drugs. The hormones are released by the action of proteolytic enzymes. Usually, however, the term prodrug is restricted to artificially synthesised molecules that are acted upon by the

Leu →D—Phe →Pro →Val →Orn   Orn→Leu →D—Phe →Pro →Phe

↑   ↓   ↑   ↓

Orn ←Val ←Pro ←D—Phe ←Leu   Val ←Tyr ←Gln ←Asn ←D—Phe

(9·7)   (9·8)

body's enzymes to release a pharmacologically active molecule. The latter may be a naturally occurring molecule or one that is purpose designed.

A pro-drug may be preferable to the drug itself for various reasons. First, it may be desirable to protect the alimentary canal from the action of the drug. Secondly, it may be desirable to protect the drug from the enzymes in the digestive system. Thirdly, it may be necessary to modify the physical properties of the drug in order that it shall be possible to direct it to the required site. For example, a hydrophilic molecule is unlikely to able to cross the blood–brain barrier and act on the brain. If the drug is incorporated into a hydrophobic molecule, however, the pro-drug may be able to reach the brain and the active component can be released by proteolysis on site. Finally, it may be possible to design a pro-drug that can only be activated by a microbial enzyme. Any possible side effects of the drug would be minimal with such a system. Although many prokaryotic enzymes have eukaryotic analogues, such a pro-drug is feasible since there are enzymes that are unique to prokaryotes. Although the concept of designing pro-drugs looks very attractive in principle, in practice there have been no remarkable successes.

9.9 Peptide antibiotics

Some antibiotics that have been derived from peptides were mentioned in Chapter 1. The biosynthesis of penicillins was discussed in Chapter 8. Many peptide antibiotics are known. Some find clinical applications but others such as gramicidin S (**9.7**), tyrocidine A (**9.8**) and polymyxins (**9.9**) are too toxic for use in humans. Cyclosporin A (Figure 1.4), however, has immunosuppressive properties and it has been used in transplant surgery for this reason rather than for its antibiotic properties. Peptide antibiotics have some non-standard structural features and these may explain in part their antibiotic properties. First, cyclic peptides are not found in animal cells. Secondly, peptide antibiotics usually contain some unusual amino acids; they may have the D configuration, be N-methylated or have other non-standard structural features. Clearly, these features are not compatible with direct ribosomal synthesis.

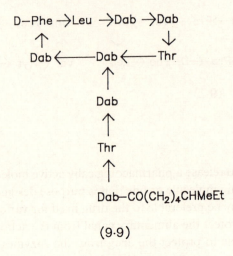

(9.9)

Dab = 2,4–Diaminobutyric acid

9.10 References

9.10.1 References cited in the text

Donadio, S., Perks, H. M., Tsuchiya, K. and White, E. H. (1985) *Biochemistry*, **24**, 2447.

Drijfhout, J. W. and Bloemhoff, W. (1991) *Int. J. Peptide Protein Res.*, **37**, 27.

Goddard, P., McMurray, J. S., Sheppard, R. C. and Emson, P. (1988) *J. Chem. Soc., Chem. Commun.*, 1025.

Groutas, W. C., Badger, R. C., Ocain, T. D., Felker, D., Frankson, J. and Theodorakis, M. (1980) *Biochem. Biophys. Res. Commun.*, **95**, 1890.

Hardie, D. G. (1991) *Biochemical Messengers*, Chapman & Hall, London.

Janda, K. D. (1994) *Proc. Natl. Acad. Sci., U. S. A.*, **91**, 10779.

Nestler, H. P., Bartlett, P. A. and Still, W. C. (1994) *J. Org. Chem.*, **59**, 4723.

Pauling, L. (1946) *Chem. Eng. News*, **24**, 1375.

Rodriguez, M., Lignon, M.-F., Galas, M. C., Fulcrand, P., Mendre, C., Aumelas, A., Laur, J. and Martinez, J. (1987) *J. Med. Chem.*, **30**, 1366.

Sasaki, Y., Murphy, W. A., Heiman, M. L., Lance, V. A. and Coy, D. H. (1987) *J. Med. Chem.*, **30**, 1162.

Tam, T. F., Spencer, R. W., Thomas, E. M., Copp, L. J. and Krantz, A. (1984) *J. Amer. Chem. Soc.*, **106**, 6849.

Wolfenden, R. (1972) *Acc. Chem. Res.*, **5**, 10.

9.10.2 References for background reading

Basava, C. and Anantharamaiah, G. M. (Eds.) (1994) *Peptides: Design, Synthesis and Biological Activity*, Birkhauser, Boston; Springer Verlag, New York.

Bloom, S. R. and Burnstock, G. (Eds.) (1991) *Peptides: A Target for New Drug Development*, IBC, London.

Dutta, A. (1993) *Small Peptides: Chemistry, Biology and Clinical Studies. Pharmacochemistry Library*, Vol. 19, Elsevier, Amsterdam.

Gallop, M. A., Barrett, R. W., Dower, W. J., Fodor, S. P. A. and Gordon, E. M. (1994) *J. Med. Chem.*, **37**, 1233. (A review on combinatorial synthesis.)

Gante, J. (1994) *Angew. Chem., Int. Ed.*, **33**, 1699. (A review on pseudo-peptide enzyme inhibitors.)

Gordon, E. M., Barrett, R. W., Dower, W. J., Fodor, S. P. A. and Gallop, M. A. (1994) *J. Med. Chem.*, **37**, 1385. (Combinatorial synthesis.)

Hider, R. C. and Barlow, D. (Eds.) (1991) *Polypeptide and Protein Drugs*, Horwood, London.

Horwell, D. C. Howson, W. and Rees, D. C. (1994) *Drug Design Discovery*, **12**, 63. (A review on peptoids.)

Voelter, W., Stoeva, S., Kaiser, T., Grubler, G., Mihelic, M., Echner, H., Haritos, A. A., Seeger, H. and Lippert, T. H. (1994) *Pure Appl. Chem.*, **66**, 2015. (Design of synthetic peptide antigens.)

Ward, D. J. (1991) *Peptide Pharmaceuticals*, Open University Press, Milton Keynes.

Wisdom, G. B. (1994) *Peptide Antigens: A Practical Approach*, IRL Press, Oxford.

Subject index